CHEMICAL ANALYSIS

CHEMICAL ANALYSIS

A SERIES OF MONOGRAPHS ON
ANALYTICAL CHEMISTRY AND ITS APPLICATIONS

VOLUME 35

WILEY-INTERSCIENCE

A Division of John Wiley & Sons, Inc., New York/London/Sydney/Toronto

Laser Raman Spectroscopy

Marvin C. Tobin

*The Perkin-Elmer Corporation
Norwalk, Connecticut*

and

*The University of Bridgeport
Bridgeport, Connecticut*

WILEY-INTERSCIENCE

Division of John Wiley & Sons, Inc., New York • London • Sydney • Toronto

Library of Congress Catalogue Card Number: 70-148511

ISBN 0-471-87550-3

Printed in the United States of America.

10 9 8 7 6 5 4 3 2 1

לזכר נשמת אבי מורי זלמן בן חיים יעקב הלוי

To the memory of my father, Sol Tobin

PREFACE

With perhaps half a dozen excellent books on Raman spectroscopy already on the market, the reader might question the need for still another one. The necessity for a new text arises from the recent developments in instrumentation, which have revolutionized Raman spectroscopy. The nearly simultaneous development of lasers, low-dark-current photomultiplier tubes, and efficient double monochromators has, at last, made the Raman effect a useful routine method. Several excellent commercial instruments embodying these components are already on the market.

This book emphasizes the experimental techniques of laser Raman spectroscopy. Only enough theory is given, in Chapter 1, to facilitate interpretation of the experimental data. The presentation is detailed enough, however, to serve as a basis for the study of more advanced texts. Group theory, normal coordinate analysis, and the quantum theory of scattering are already treated in so many places that it did not seem profitable to present another detailed exposition. A chapter on characteristic frequencies in the infrared and Raman spectrum treats the more important chemical groupings. For details of these subjects, the reader is referred to the General References. By the same token, some topics, although of great scientific interest, are barely mentioned. Among these are the electronic Raman effect, the second-order Raman effect, and details of the analysis of rotational spectra. Since this book is aimed primarily at chemists interested in chemical analysis and molecular structure determination, such topics seemed to be of limited interest. On the other hand, the Raman spectra of single crystals are treated in some detail.

It is assumed that the reader is familiar with the elements of classical and quantum mechanics and has at least some knowledge of the use of infrared spectroscopy for structure determination.

As the book was being written, it became apparent that a good part of the text dealt with spectrometers, principles of optics, data processing, and the like. The chemist using a good commercial Raman spectrometer to record the spectra of high-quality samples may skip this part of the book. However, if a research worker wishes to assemble his own Raman spectrometer, to build specialized sampling accessories, or to examine very turbid or fluorescent samples, he will find the information given valuable. It might be mentioned that exactly the same principles hold for laser-excited fluorescence spectroscopy, and, to some degree, for any kind of absorption or emission spectroscopy. The reader will note that in several places, principles are illustrated with worked-out examples. I have learned, in ten years of teaching physics to students, that one worked-out example is worth a thousand words.

MARVIN C. TOBIN

Bridgeport, Connecticut
November 1970

CONTENTS

1

THE NATURE OF THE RAMAN EFFECT

1.1 DESCRIPTION

INTRODUCTION

The Raman effect is one member of a wide class of light-scattering phenomena. These all have their origin in the fact that a collimated beam of light passed through a nonabsorbing medium is invariably found to be attenuated by the medium. The energy lost from the primary beam is not degraded to heat. Instead, it is scattered into the 4π steradians of solid angle, around the sample. The most familiar everyday example is the scattering of light by dust particles in a sunbeam. It is useful at this point to distinguish between colloidal and molecular scattering. The overall features of light scattering are markedly different, if the sample consists of one phase finely dispersed in another, from the case where the sample is uniform. Only the latter case is of interest to us in this book. The overall features of light scattering are most easily described if it is assumed that the illuminating beam is highly collimated and highly monochromatic.

Suppose, now, that some portion of the light scattered from a sample is passed through a spectrometer. The bulk of the scattered energy will be found at the frequency of the incident beam. However, a small portion of the scattered energy will be found at definite frequencies above and below that of the incident beam. If the frequency of the incident beam is arbitrarily taken as zero, then the displaced frequencies will be found to correspond to some or all of the normal frequencies of the sample. This modulation of the frequency of the incident beam is the Raman effect. To put matters into focus, suppose that the incident light has a frequency of 19,436 cm^{-1}. If the scattering material contains a carbonyl group, then some of the scattered radiation will be found to have a frequency near 17,786 cm^{-1}. A much smaller amount will be found to have a frequency near 21,086 cm^{-1}. The difference between the frequencies of the incident and shifted radiations, 1650 cm^{-1}, is the Raman frequency associated with the particular carbonyl group in the sample. The agency modulating the frequency of the incident light turns out to be the normal frequencies of the sample. These may be molecular rotational or vibrational modes of a gas, lattice modes of a crystal, vibrational modes of a glass, and so on. Since both the Raman spectrum and the infrared absorption spectrum of a material arise from the same basic source, they yield somewhat the same information. The fundamentally different mechanisms of the two phenomena, however, produce major differences in the two spectra.

It should be emphasized that the numbers derived from the Raman spectra are the differences between incident and shifted frequencies. For this reason, the shapes and positions of the Raman bands do not depend on the frequency of the exciting radiation. (The intensity of the scattered light is strongly dependent on this quantity.)

One should be careful to distinguish between fluorescence and scattering phenomena. In fluorescence, the system absorbs radiation and is excited to a higher electronic state. After some 10^{-8} sec, it reemits radiation. Raman or other scattering processes involve no absorption of radiation. The scattering system is never in an excited electronic state. The whole process takes place in less than 10^{-12} sec.

GROSS EXPERIMENTAL METHOD

Chapter 2 is devoted to a description of modern methods of recording Raman spectra. The present discussion is solely for the purpose of providing historical background and a background for theory. Figure 1.1 shows a block diagram of a Raman spectrograph. A sample is illuminated by a suitably filtered source of monochromatic light. The scattered light (generally viewed at 90° to the direction of the incident light) is picked up by a series of lenses and passed through the slit of a spectrograph. The spectrum is then either recorded on a photographic plate, or scanned with a photomultiplier tube. In either case, the spectrum is exhibited as a plot of intensity versus frequency. Up to about 1965, the most commonly used source of exciting radiation was a helical mercury arc, surrounding a cylindrical sample tube lying along its axis. The scattered radiation was viewed through a window on the bottom of the sample tube. Use of a mercury arc is attended by numerous difficulties. The region from 4000 to 6000 Å contains some eight strong mercury lines plus a continuous, diffuse background. To record a satisfactory spectrum, it is necessary to isolate a single one of the strong lines and to filter out the continuous background. The problem of doing this and still delivering an adequate amount of radiation to the sample has never been completely solved. Nevertheless, chemical filters for the mercury arc radiation were developed which made it possible to record satisfactory Raman spectra of many materials in a reasonable time. Until very recently, most Raman spectra were recorded on a photographic plate. One commercial Raman spectrometer (the Cary Model 81) using photoelectric detection, has been on the market since the 1950s. Possibly due to cost, its use did not become widespread. This was in spite of the fact that it yielded spectra much

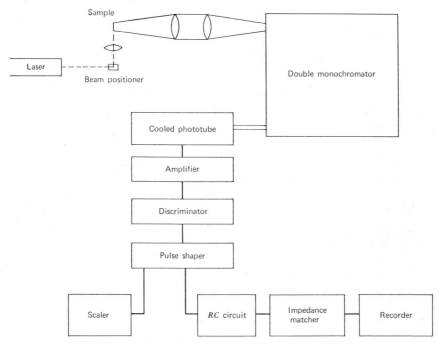

Fig. 1.1 Block diagram of a laser Raman spectrometer.

superior to those obtainable from a spectrograph with photographic record-ing. Figure 1.2 shows a typical photographic Raman spectrum and a spec-trum recorded with modern instrumentation.

Although the experimental methods just described sound (and are) perfectly straightforward, they are accompanied by numerous difficulties. The most commonly used mercury line is at 4358 Å. If the sample is colored more red than a pale yellow, this line cannot be used.[1] The green line at 5461 Å or the yellow doublet at 5770–5790 Å were sometimes used to obtain Raman spectra of colored samples. Because of their relative feeble-ness and because of the lower sensitivities of photographic plates or photo-electric detectors in the red, these lines were seldom used. Helium, cad-mium and other lamps were used with some success to excite the spectra of colored samples.[2]

Far more serious difficulties arise if the sample is turbid or fluorescent or both. It will be recalled that most of the incident radiation scattered by a sample is at the frequency of the incident radiation. Only a small fraction, generally about one part in 10^{-5}, is shifted in frequency. If the sample is turbid, it scatters far more of the incident light in proportion to the fre-quency-shifted Raman light. Once the light at the incident frequency is

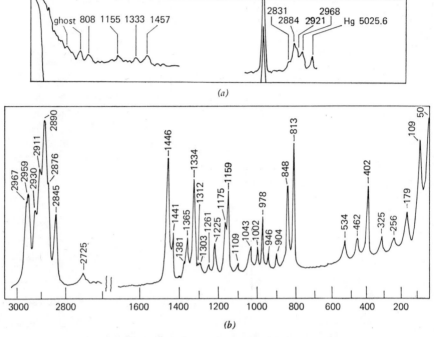

(a)

(b)

Fig. 1.2 (*a*) A densitometer tracing of the Raman spectrum of isotactic polypropylene, excited with Hg 4358 Å and recorded photographically. (*b*) A modern Raman spectrum of isotactic polypropylene, excited with He-Ne laser excitation.

inside the spectrograph, it tends to be scattered from the optical surfaces, and darken the photographic plate. (This difficulty was less severe with the Cary Model 81 Raman spectrometer, which was equipped with a double monochromator for photoelectric detection.) Fluorescence produces a similar blackening of a photographic plate, since it consists of a relatively smooth, continuous background underlying the Raman bands.

A final difficulty in recording Raman spectra with pre-1965 instrumentation is detector noise. A photographic plate is grainy and invariably has some background haze, when developed. This makes it hard to detect weak Raman bands. Again, only in the last few years have photomultiplier tubes been developed with extremely low dark current. Shot noise on phototube dark current makes it difficult to detect weak Raman bands for the same reason as does haze in a photographic plate.

As a result of the foregoing, most of the thousands of Raman spectra reported in the literature were obtained from colorless, stable liquids. Some work was done with crystals, crystal powders, gases and the like, but only a very few laboratories undertook these studies. Although attempts were made to use Raman spectroscopy for routine quantitative or qualitative analysis, the experimental difficulties kept it from taking hold.

One might ask, at this point, if Raman spectroscopy was so difficult experimentally, why did people not just use the infrared spectrum? The answer is that much of the information contained in the Raman spectrum does not appear in the infrared spectrum. Some normal vibrations of a system may appear in the infrared spectrum only, in the Raman spectrum only, in both or in neither. Certain group frequencies, such as ethylenic C=C stretching bands, are strong in the Raman spectrum, weak in the infrared spectrum. Water is an exceptionally weak scatterer in the Raman effect, but an exceptionally strong absorber in the infrared spectrum. Raman spectroscopy is thus the preferred method if one wishes to study aqueous solutions.

Fortunately, at the time of the writing of this book, all of the experimental difficulties with Raman spectroscopy have been overcome, with the one exception of sample fluorescence. Even this is, in most cases, manageable. The experimental techniques given in Chapter 2 will deal almost exclusively with laser excitation and photoelectric detection of Raman spectra. These have almost completely displaced the earlier techniques. The only allusions to mercury arc excitation will be for historical or for comparison purposes.

1.2 INTRODUCTORY THEORY

GENERAL THEORY

The existence of the Raman effect was predicted, some time before its experimental demonstration, on the basis of the analysis starting with equation 1.2.1. This section is intended to give the reader enough understanding of the theory of the Raman effect to apply it to the interpretation of experimental data. Detailed derivations are given in a number of standard texts.

A molecule placed in an electromagnetic field has its charge distribution periodically disturbed by the field.[R1] The resultant induced, alternating dipole moment acts as a source of radiation and gives rise to the entire class of light scattering phenomena described in Section 1.1.

The alternating dipole moment is generally expressed as the dipole moment per unit volume, the polarization. In the case of interest here, the polarization is proportional to the inducing field

$$\bar{P} = \alpha \bar{E} \tag{1.2.1}$$

The constant α, the polarizability, is of central importance in the theory of the Raman effect. Although the description of the details of the Raman effect requires quantum theory, the existence of the effect is easily predicted from classical electromagnetic theory.

The inducing electromagnetic field is given by

$$\bar{E} = \bar{E}_0 \cos 2\pi\nu t \qquad (1.2.2)$$

We then expect a polarization

$$\bar{P} = \alpha \bar{E}_0 \cos 2\pi\nu t \qquad (1.2.3)$$

The polarizability, α, consists of two parts. The first is a constant, α_0, which represents the static polarizability. The second is a sum of terms having the periodic time dependence of the normal frequencies of the system under consideration. (Normal frequencies are discussed in Section 1.3). The polarizability may then be written as

$$\alpha = \alpha_0 + \sum \alpha_n \cos 2\pi\nu_n t \qquad (1.2.4)$$

The normal frequencies ν_n may be rotational or vibrational frequencies of a molecule, lattice frequencies of a crystal or even, in Brilliouin scattering, acoustic frequencies of a solid. If (1.2.2) and (1.2.4) are set into (1.2.1), we get

$$\bar{P} = \bar{E}_0 \alpha_0 \cos 2\pi\nu t + \bar{E}_0 \sum \alpha_n \cos 2\pi\nu t \cos 2\pi\nu_n t$$
$$= \bar{E}_0 \alpha_0 \cos 2\pi\nu t + \tfrac{1}{2}\bar{E}_0 \sum \alpha_n \{\cos 2\pi(\nu - \nu_n)t + \cos 2\pi(\nu + \nu_n)t\}$$
$$(1.2.5)$$

Equation 1.2.5 correctly predicts the major qualitative features of the Raman effect. First, there is the leading term, which represents the component of the polarization having the frequency of the exciting field. This accounts for Rayleigh scattering. Second, each variable component of the polarizability, α_n, gives rise to components of the polarization having frequencies $(\nu + \nu_n)$ and $(\nu - \nu_n)$. These account for the Stokes and anti-Stokes Raman bands.

Actually, equation 1.2.1 is too restrictive. It predicts that the polarization will have the same direction relative to a fixed set of coordinate axes as the electric field. In fact, the polarizability is a tensor. As will be seen, the angular dependence of Rayleigh and Raman scattering as well as the polarization of the scattered light, are consequences of the tensor properties of α. The relation between $\bar{P}$ and $\bar{E}$ then reads

$$P_x = \alpha_{xx} E_x + \alpha_{xy} E_y + \alpha_{xz} E_z$$
$$P_y = \alpha_{yx} E_x + \alpha_{yy} E_y + \alpha_{yz} E_z \qquad (1.2.6)$$
$$P_z = \alpha_{zx} E_x + \alpha_{zy} E_y + \alpha_{zz} E_z$$

where the α_{ij} are the component of the polarizability tensor. Equation 1.2.6 may be written in matrix notation.

$$\begin{vmatrix} P_x \\ P_y \\ P_z \end{vmatrix} = \begin{vmatrix} \alpha_{xx} \alpha_{xy} \alpha_{xz} \\ \alpha_{yx} \alpha_{yy} \alpha_{yz} \\ \alpha_{zx} \alpha_{zy} \alpha_{zz} \end{vmatrix} \begin{vmatrix} E_x \\ E_y \\ E_z \end{vmatrix} \qquad (1.2.7)$$

(See Section 1.4 for an explanation of matrix notation.) It should be mentioned that if one introduces a coordinate system, the relation of the vectors $\bar{E}$ and $\bar{P}$ has the tensor property of being independent of the coordinate system, although the components of $\bar{P}$, α, and $\bar{E}$ may change. This will prove to be important in connection with the discussion of crystals. From classical electromagnetic theory, the energy scattered per unit time is given by[R1]

$$I = \frac{2\overline{\bar{P}^2}}{3c^2} \tag{1.2.8}$$

Therefore, the scattered radiation is predicted to contain frequency components at $v \pm v'$.

The classical description given correctly predicts the existence of the Raman effect and properly describes its dependence on the frequency of the exciting radiation. It does not predict the fact that some normal frequencies do not give rise to Raman scattering. This and other fine details must be deduced from a quantum mechanical analysis.

The quantum mechanical description of the Raman effect starts from a calculation of the dipole moment associated with the transition between two states, a and b, when a molecule is perturbed by an electromagnetic field of frequency v.

This quantity turns out to have the properties necessary to describe the Raman effect. It is given (after a lengthy calculation) by[R1]

$$
\begin{aligned}
|\bar{R}_{ab}| = {}& 2(a|\bar{R}|b) \cos 2\pi v_{ab} t \\
& + \sum_j \left| \frac{v_{aj}}{v_{aj}+v} - \frac{v_{jb}}{v_{jb}-v} \right| (a|\bar{R}|j)(j|\bar{R}|b) \\
& \qquad \cdot \frac{E^0}{hv} \sin 2\pi(v+v_{ab})t \\
& + \sum_j \left| \frac{v_{bj}}{v_{bj}+v} - \frac{v_{ja}}{v_{ja}-v} \right| (a|\bar{R}|j)(j|\bar{R}|b) \\
& \qquad \cdot \frac{E^0}{hv} \sin 2\pi(v-v_{ab})t
\end{aligned}
\tag{1.2.9}
$$

Equation (1.2.9) requires some explanation. The quantity $\bar{E}^0$ is the intensity of the incident electromagnetic field and v its frequency. The quantities $(a|\bar{R}|j)$, and so on, are matrix elements of $\bar{R} = e(x\bar{i} + y\bar{j} + z\bar{k})$ between the various eigenstates of the molecule. That is,

$$(a|\bar{R}|j) = \bar{i} \int u_a^* x u_j \, dv + \bar{j} \int u_a^* y u_j \, dv + \bar{k} \int u_a^* z u_j \, dv \tag{1.2.10}$$

The ν_{aj}, ν_{ab}, and ν_{jb} are the frequencies of the transition between the states given by the subscript letters. Suppose, now, that the state a is the ground state of the molecule and b one of the vibrational states. Then $|\bar{R}_{ab}|$ is seen to contain frequency components in the sine terms which are the sum and difference of the perturbing and molecular vibrational frequencies. This is, again what is needed to have a Raman effect. Since there is a summation over j, all of the possible vibrational, rotational and electronic states of the molecule may contribute to $|\bar{R}_{ab}|$. Whenever ν, the perturbing frequency, approaches one of the ν_{jb}, $|\bar{R}_{ab}|$ will become very large. (It will not become infinite, since this treatment neglects a damping term in the resonance denominators.[R2])

The energy radiated per second by a system, put in an electromagnetic field, into a Raman transition, is given by

$$I = \frac{64\pi^4(\nu \pm \nu_{ab})^4}{3c^3} |\bar{R}_{ab}|^2 \tag{1.2.11}$$

Therefore, our $|\bar{R}_{ab}|$ does, indeed, describe a Raman effect.

It might appear that there is no relation between (1.2.7) and (1.2.9). Actually, both equations have the same form. To see this, consider a term like $(a|\bar{R}|j)(j|\bar{R}|b) \cdot \bar{E}^0 = \bar{X}$, in equation 1.2.9. Let $(a|\bar{R}|j) = \bar{a}$ and let $(j|\bar{R}|b) = \bar{b}$. We calculate the components of $\bar{X}$. To start, we multiply out the dot product.

$$\bar{X} = (b_x E_x^0 + b_y E_y^0 + b_z E_z^0)\bar{a} \tag{1.2.12}$$

Taking the components of $\bar{X}$, we find

$$X_x + b_x E_x^0 a_x + b_y E_y^0 a_x + b_z E_z^0 a_x = a_x b_x E_x^0 + a_x b_y E_y^0 + a_x b_z E_z^0$$
$$X_y = b_x E_x^0 a_y + b_y E_y^0 a_y + b_z E_z^0 a_y = a_y b_x E_x^0 + a_y b_y E_y^0 + a_y b_z E_z^0$$
$$X_z = b_x E_x^0 a_z + b_y E_y^0 a_z + b_z E_z^0 a_z = a_z b_x E_x^0 + a_z b_y E_y^0 + a_z b_z E_z^0$$

$$\tag{1.2.13}$$

In matrix notation,

$$\begin{vmatrix} X_x \\ X_y \\ X_z \end{vmatrix} = \begin{vmatrix} a_x b_x & a_x b_y & a_x b_z \\ a_y b_x & a_y b_y & a_y b_z \\ a_z b_x & a_z b_y & a_z b_z \end{vmatrix} \begin{vmatrix} E_x^0 \\ E_y^0 \\ E_z^0 \end{vmatrix} \tag{1.2.14}$$

which has the form of equation 1.2.7.

The last two terms in equation 1.2.9 have the form of the quantity $\bar{X}$. It is clear from (1.2.7), (1.2.9), and (1.2.14) that $|\bar{R}_{ab}|$ is the quantum mechanical polarization. (The first term in (1.2.9) represents radiation of quanta at the frequency of the molecular vibration, ν_{ab}, and has no classical equivalent.) Equation 1.2.14 controls the properties of the Raman effect. Consider, for example, the tensor element

$$a_x b_y = \int u_a^* x u_j \, dv \int u_j^* y u_b \, dv \tag{1.2.15}$$

If any pair of these integrals differs from zero for any u_j and for any product of x, y, and z then the Raman effect can occur for the transition $a \rightarrow b$. It is, of course, not practical to compute these integrals directly and equation 1.2.9 is rarely used for calculation. It turns out that the symmetry properties of the system determines whether or not the tensor elements like 1.2.15 are zero. Calculation of the integral is, therefore, unnecessary if one wishes to know only whether or not the transition will occur. Generally, a is the ground state of the system and b a rotational or vibrational energy level. Electronic Raman effects in which b is an electronic level are known but these will not be discussed in this book.

ANGULAR DEPENDENCE[3]

Further progress requires consideration of the directional properties of light scattering. These are consequences of the fact that the polarizability is a tensor. Therefore, although the discussion given here is applied to Raman scattering in particular, it is applicable to Rayleigh and Brillouin scattering, as well. Furthermore, in view of equations 1.2.7 and 1.2.14, the results are equally valid for the classical and the quantum mechanical treatments.

Consider a molecule which has a fixed orientation relative to a set of coordinate axes. It is illuminated with radiation directed along the y axis, the scattered radiation being observed in the xy plane. If the incident radiation is polarized along the z axis, only E_z^0 contributes to the scattering. (Fig. 1.3a).

If the angle of observation, $\theta = 90°$, we have*

$$\begin{aligned} P_x &= 0 \\ P_y &= \alpha_{yz} E_z^0 \\ P_z &= \alpha_{zz} E_z^0 \end{aligned} \qquad (1.2.16a)$$

If the angle θ is zero, we have

$$\begin{aligned} P_x &= \alpha_{xz} E_x^0 \\ P_y &= 0 \\ P_z &= \alpha_{zz} E_z^0 \end{aligned} \qquad (1.2.16b)$$

For intermediate angles, we have

$$\begin{aligned} P_x &= a_{xz} E_z^0 \cos \theta \\ P_y &= \alpha_{yz} E_z^0 \sin \theta \\ P_z &= \alpha_{zz} E_z^0 \end{aligned} \qquad (1.2.17)$$

*P_x is, of course, not zero, but $\alpha_{zz} E_z^0$. Since, however, the direction of observation is along the x axis and P_x is directed along this axis, the observer cannot see P_x. The same is true for P_y in (1.2.16b).

The argument is equally valid whether $\bar{P}$ is the classical polarization or the quantum mechanical $|\bar{R}_{ab}|$. For the polarization of incident light shown in Figure 1.3a, the total energy scattered at angle θ is from equation 1.2.17 and from equation 1.2.8 or 1.2.11, proportional to

$$P_x^2 + P_y^2 + P_z^2 = (\alpha_{zz}^2 + \alpha_{xz}^2 \cos^2\theta + \alpha_{yz}^2 \sin^2\theta)(E_z^0)^2 \quad (1.2.18)$$

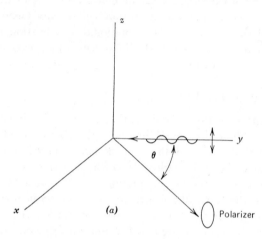

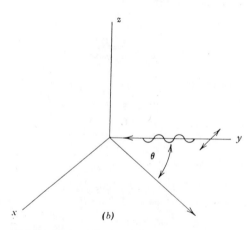

Fig. 1.3 Directional dependence of Raman scattering. The laser beam is along the y axis, with its electric vector as shown. The direction of observation is the skew direction in the xy plane. (a) Situation $I(\perp)$. Polarizer E is in the z direction for $I(\perp)(\text{obs}\perp)$, in the xy plane for $I(\perp)(\text{obs}\|)$. (b) Situation $I(\|)$.

In Figure 1.3b, the incident light has its polarization vector along the x-axis. A little consideration shows that, in this case, the total scattered intensity is proportional to

$$P_x^2 + P_y^2 + P_z^2 = (\alpha_{zx}^2 + \alpha_{yx}^2 \sin^2 \theta + \alpha_{xx}^2 \cos^2 \theta)(E_x^0)^2 \qquad (1.2.19)$$

As a function of θ, equations 1.2.18 and 1.2.19 represent the sum of a circle and an ellipse. It should be emphasized that equations 1.2.16 through 1.2.19 are predicated on the molecule maintaining a fixed orientation relative to the coordinate axes. If this orientation is changed, the equations will remain valid, but the components α_{ij} will change. It is possible to show that there always exists a coordinate system (i.e., a particular orientation of the molecule relative to the axes) for which equation 1.2.6 takes the form

$$\begin{aligned} P_x &= \alpha_1 E_x \\ P_y &= \alpha_2 E_y \\ P_z &= \alpha_3 E_z \end{aligned} \qquad (1.2.20)$$

If the molecule has elements of symmetry, this orientation is easily picked out. Equation 1.2.20 is particularly important in the Raman spectroscopy of crystals.

Let us now return to equation 1.2.17 and suppose that the Raman radiation is being observed, through a polarizer, along the x axis, that is, at $0 = 90°$. Then, $P_y^2 = \alpha_{yz}^2(E_z^0)^2$ and $P_z^2 = a_{zz}^2(E_z^0)^2$ can be measured separately by setting the polarizer E vector parallel to the y and z axis, respectively. It follows that

$$\frac{P_y^2}{P_z^2} = \frac{\alpha_{yz}^2}{\alpha_{zz}^2} \qquad (1.2.21)$$

This is to say that, unless $\alpha_{yz} = \alpha_{zz}$, the scattered radiation is partly polarized. It should be clear at this point that where the scattering system can be put into a definite orientation, as is the case with a crystal, the individual components of the polarizability, α_{ij}, can be measured by measuring the intensity of the scattered radiation. For gases or liquids, only averages over the polarizability elements can be measured. The averaging is carried out by assuming that, initially, the molecule is in such an orientation that equation 1.2.20 holds, and then averaging the polarizabilities over all possible angles.[R3] If we define

$$\alpha^2 = \tfrac{1}{3}(\alpha_1 + \alpha_2 + \alpha_3)$$

and

$$\beta^2 = \tfrac{1}{2}[(\alpha_1 - \alpha_2)^2 + (\alpha_2 - \alpha_3)^2 + (\alpha_3 - \alpha_1)^2] \qquad (1.2.22)$$

we find, for the situation of Figure 1.3a, that

$$I(\perp)(\text{obs } \perp) \propto \frac{45\alpha^2 + 4\beta^2}{45} \qquad (1.2.23)$$

and

$$I(\perp)(\text{obs } \|) \propto \frac{\beta^2}{15} \qquad (1.2.24)$$

independent of angle of observation and with the proportionality constant the same for both equations. The ratio of the last two quantities is defined as the *depolarization* ratio of the Raman band

$$\rho = \frac{3\beta^2}{45\alpha^2 + 4\beta^2} \qquad (1.2.25)$$

The value of ρ may vary from 0 for large α to 0.75 for large β.

The total scattering itensity, I for the case of Figure 1.3a is given by

$$I \propto \frac{45\alpha^2 + 7\beta^2}{45} \qquad (1.2.26)$$

independent of angle of observation.

For the configuration of Figure 1.3b, where the plane or polarization of the incident light is in the plane of observation, the intensity of the scattered Raman radiation does depend on the direction of observation.

The *total* light scattered, in this case, is given by

$$I \propto [6\beta^2 + (45\alpha^2 + \beta^2) \cos^2 \theta] \propto [6 + \left(\frac{3 - 3\rho}{\rho}\right) \cos^2 \theta] \qquad (1.2.27)$$

where ρ is given by equation 1.2.25. The angular dependence of the scattering into any Raman band is thus known, once the depolarization ratio is known.

The depolarization ratio, as will be seen below (p. 68) is an extremely important characteristic of a Raman band. The angular dependence is of less interest, since it may be predicted from the depolarization ratio.

THE POLARIZABILITY THEORY[R2]

While the foregoing formulation of the Raman effect is the fundamental one, it is too complicated to lend itself to general application. Placzek has developed a theory which, while more restrictive, is much more useful in practice than the general theory. Provided that the frequency of the exciting radiation is far from any absorption band of the sample, and greater than the frequencies of the Raman transitions, the restricted theory holds well.

In these circumstances, the elements of the polarizability tensor depend only on the coordinates of the nuclei and not on the electron coordinates.

We start by asking, what is the quantum mechanical expectation value of one of the matrix elements, $[\alpha_{ij}]^{ab}$ associated with the transition $a \to b$? From quantum mechanics, some operator, α_{ij}, will correspond to the polarizability and

$$[\alpha_{ij}]^{ab} = \int u_b^*(q)\alpha_{ij}u_a(q)\,dq \qquad (1.2.28)$$

Equation 1.2.28 requires some explanation. The symbol q, is a shorthand notation for the $3N$ normal coordinates of the system. The wave function u is presumed to be expressed in terms of these coordinates. Furthermore, the wave functions themselves are products of wave functions each of which depends on only one normal coordinate. This is to say that $u_a = \Pi_{n=1}^{3N} w_n(q_n, v_n)$, with a similar expression for u_b. The quantity v_n is a quantum number associated with the wave function w_n.

The operator α_{ij} is now assumed to be a function of the normal coordinates and is expanded into a Taylor series in these coordinates. We get

$$\alpha_{ij}(q) = (\alpha_{ij})_0 + \sum_{n=1}^{3N} \left(\frac{\partial \alpha_{ij}}{\partial q_n}\right)_0 q_n + \cdots \qquad (1.2.29)$$

We now substitute (1.2.29) and the expressions for u_a and u_b into (1.2.28). We simplify the notation by dropping the subscripts ij and writing $w_n(q_n, v_n)$ simply as $w(v_n)$. This gives us

$$[\alpha_{ij}]^{ab} = (\alpha_0) \prod_{n=1}^{3N} \int w(v_n')w(v_n)\,dq_n$$
$$+ \sum_{n=1}^{3N} \left(\frac{\partial \alpha}{\partial q_n}\right)_0 \prod_{m \neq n} \int w(v_m')w(v_m)\,dq_m \cdot \int w(v_n')q_n w(v_n)\,dq_n \qquad (1.2.30a)$$

This reduces to

$$[\alpha_{ij}]^{ab} = \alpha_0 + \sum_{n=1}^{3N} \left(\frac{\partial \alpha}{\partial q_n}\right) \int w(v_n')q_n w(v_n)\,dq_n \qquad (1.2.30b)$$

Equation 1.2.30b is derived from equation 1.2.30a as follows. The operator α_0 corresponds to the static polarizability and is a constant. The integrals which multiply it equal zero if $v_n' \neq v_n$ and equal one if $v_n' = v_n$. We are left with a constant, α_0, in (1.2.30b). This constant corresponds to Rayleigh scattering, as in the classical treatment. By the same token, the integrals $\int w(v_m')w(v_m)\,dq_m$ equal zero if $v_m' \neq v_m$ and equal one if $v_m' = v_m$. The product reduces to one and we are left with (1.2.30b). If one uses the harmonic oscillator approximation, the integrals in (1.2.30b) can be evaluated. If one introduces the Boltzmann expression for the population density of the energy levels an expression for the scattering intensity in terms of $(\partial\alpha/\partial q)_0$ may be derived. The total power scattered per molecule into a given Stokes Raman band, when the molecule is illuminated with I_0 watts cm^{-2} is[R4]

$$I = \frac{K}{M} \frac{(\nu_0 - \nu)^4 g(45\alpha^2 + 7\gamma^2)}{\nu(1 - \exp(-h\nu/kT))} I_0 \qquad (1.2.31)$$

Here g is the degeneracy of the molecular level giving rise to the Raman scattering, ν_0 the exciting frequency, M the molecular weight, K a known constant, and ν the Raman frequency. The incident light may have arbitrary polarization, since the intensity refers to scattering into 4π steradians. The quantities α and γ are given by

$$\alpha = \tfrac{1}{3}\left[\left(\frac{\partial\alpha_x}{\partial q}\right)_0 + \left(\frac{\partial\alpha_y}{\partial q}\right)_0 + \left(\frac{\partial\alpha_z}{\partial q}\right)_0\right] \qquad (1.2.32)$$

and

$$\gamma^2 = \tfrac{1}{2}\left[\left(\frac{\partial\alpha_x}{\partial q} - \frac{\partial\alpha_y}{\partial q}\right)^2 + \left(\frac{\partial\alpha_y}{\partial q} - \frac{\partial\alpha_z}{\partial q}\right)^2 + \left(\frac{\partial\alpha_z}{\partial q} - \frac{\partial\alpha_x}{\partial q}\right)^2\right] \qquad (1.2.33)$$

It should be emphasized that equation 1.2.31 is valid only for gases, and to a limited degree, for non-polar liquids, since complete randomness of orientation was assumed in the derivation.

From the discussion of the preceding section, it should be clear that a measurement of the absolute intensity and the depolarization ratio of a Raman band suffice to determine α and γ of (1.2.32) and (1.2.33) for the band. Unfortunately, these quantities have little direct meaning in terms of chemical bond properties. By making some rather drastic assumptions, it is possible to use equations 1.2.31, 1.2.32 and 1.2.33 to derive bond parameters. A thorough discussion is given by Hester.[R4]

1.3 VIBRATION AND ROTATION OF MOLECULES

Let us consider a molecule fixed relative to a system of coordinate axes. Then, let us suppose that each atom is given a small, arbitrary displacement. A little reflection shows that if there are n atoms in the molecule, $3n$ numbers must be specified to fix the new positions of the atoms. If the displacement is small, the force acting on the jth atom will be, assuming simple harmonic motion

$$\begin{aligned}
F_x^j &= k_{xx}^{j1}x_1 - k_{xy}^{j1}y_1 - k_{xz}^{j1}z_1 \cdots - k_{xz}^{jn}z_n \\
F_y^j &= k_{yx}^{j1}x_1 - k_{yy}^{j1}y_1 - k_{yz}^{j1}z_1 \cdots - k_{yz}^{jn}z_n \\
F_z^j &= k_{zx}^{j1}x_1 - k_{zy}^{j1}y_1 - k_{zz}^{j1}z_1 \cdots - k_{zz}^{jn}z_n
\end{aligned} \qquad (1.3.1)$$

Since there are n atoms, $(j = 1, 2, \ldots, n)$, there will be a total of $3n$ equations. If the molecule is undergoing simple harmonic vibrations, then

$$F_x^j = 4\pi^2\nu^2 m_j x_j$$
$$F_y^j = 4\pi^2\nu^2 m_j y_j, \text{ etc.}$$
(1.3.2)

The condition that the set of equations obtained by setting (1.3.2) into (1.3.1) have a solution is that the $3n \times 3n$ determinant of the coefficients of the displacement coordinates x_j, y_j, z_j equal zero. This is to say that

$$\begin{vmatrix} k_{xx}^{11} - 4\pi^2\nu^2 m_1 & k_{xy}^{11} \cdots \cdots \cdots \cdots \\ k_{yx}^{11} & k_{yy}^{11} - 4\pi^2\nu^2 m_1 \cdots \\ & \vdots \end{vmatrix} = 0$$
(1.3.3)

Provided that the coefficients k_{lm}^{ij} are all known, the secular determinant (1.3.3) can be solved for the $3n$ values of ν^2. This gives the $3n$ normal frequencies of vibration of the molecule. The problem can, in principle be simplified. It can be shown that the problem can be set up in terms of linear combinations of the original displacement coordinates. If these are properly chosen, the secular determinant in terms of these coordinates is diagonal. This is to say that all off-diagonal elements are zero. The linear combinations of the displacement coordinates, in terms of which the secular determinant is diagonal, are called normal coordinates. The diagonal elements will all have the form

$$K_i - L_i \nu_i^2$$
(1.3.4)

Since the secular determinant equals zero, each of the elements (1.3.4) must equal zero, and the normal frequencies can be calculated directly. Unfortunately, calculation of the normal coordinates requires a prior knowledge of the normal frequencies, so that this is not a useful computational approach. It does tell us however, that the general discussion of molecular vibrations should be couched in terms of normal coordinates. Fortunately, linear combinations of the displacement coordinates, which reflect the molecular symmetry (see Section 1.4) can be found without a knowledge of the normal frequencies. In terms of these coordinates, the secular determinant factors into blocks along the diagonal. These are of smaller order than the secular determinant and can be solved individually for the normal frequencies. These can then be used to combine the symmetry coordinates to form normal coordinates.

One might think from the preceding discussion that the normal vibrations of molecules can be calculated from first principles. This would be so, were the constants k_{lm}^{ij} known. In general they are not.

The connection between the classical vibration problem and molecular spectra is made by calculating the classical normal coordinates for a molecule and setting up the quantum mechanical harmonic oscillator problem in terms of these coordinates. Schrödinger's equation is found to factor into $3n$ equations. The resultant energy levels lead to the observed infrared and Raman frequencies and the wave functions thus obtained are found to govern the observed intensities. One may look on the classical normal coordinates as simply the proper coordinates to use in order to factor the quantum mechanical Schrödinger's equation. On the other hand, many of the features of molecular vibrations are derivable from the classical treatment alone.

Returning to equation (1.3.1), suppose that the secular determinant has been solved and the $3n$ normal frequencies are known. If one of these normal frequencies is set into the $3n$ equations (1.3.1), by using (1.3.2), one gets a set of $3n$ equations in the $3n$ displacement coordinates. If any one of these coordinates is assigned an arbitrary value, the equations can be solved to give the remaining displacement coordinates in terms of it. Associated with each atom, for the normal frequency chosen, will be a vector displacement. The set of vector displacements for each frequency is called the "normal mode of vibration" for this frequency.

The discussion given so far is perfectly general and is valid whether we are talking about a single small molecule or a giant "molecule" in the form of a polymer chain or a crystal. Actually, some of the $3n$ normal frequencies of an n-atomic system will equal zero. For an ordinary molecule, if we wish to consider vibrations only, we wish to fix the center of mass and to forbid rotation of the molecule as a whole. This puts six constraints on the $3n$ cartesian coordinates, leaving $3n - 6$ vibrational normal frequencies. It is not hard to show that for a linear molecule, there are $3n - 5$ normal modes, for a polymer chain $3n - 4$ and for a crystal, $3n - 3$. In the latter two cases, n is the total number of atoms in the chain or crystal, not just the number in a unit cell.

Equation 1.3.3 can be transformed so that the force constants, k, are characteristic of bond stretching, angle bending, out-of-plane bending, and so on, motions of chemical groupings. The constants along the diagonal, then refer to pure distortions, the off-diagonal terms to interactions between these. Either the original Cartesian coordinates, or these "internal coordinates" may be further combined to take advantage of molecular symmetry. This is discussed in the next section. The "internal" coordinates are not normal coordinates and do not help factor the secular equation. The normal coordinates may, however, be written as linear combinations of internal coordinates.

1.4   SYMMETRY AND SELECTION RULES

It was stated earlier that not all of the normal modes of a system would give rise to Raman scattering or to infrared absorption. This is not to say that the spectroscopically inactive normal frequencies have no physical effect. A normal frequency which has no spectroscopic activity will still contribute to the specific heat or to the broadening of X-ray diffraction spots. The spectroscopic activity of a given normal mode is controlled by symmetry considerations. These are all based on a fundamental principle of physics: the measured value of a physical quantity must be independent of the coordinate system in which it is measured. Let us suppose for example that one of the matrix elements discussed in Section 1.2 changes sign when the coordinates (x, y, z) in the integral are changed to $(-x, -y, -z)$. This would violate the principle just stated, so that this matrix element must have the value zero. Symmetry considerations are formally treated in the branch of mathematics called group theory.[R3] The section will briefly describe group theory and its application to molecular spectroscopy. For the most part, theorems will be stated without proof, since there are numerous good texts on group theory which present such proofs.[R1, R6]

A group consists of some operation and a set of elements. As an example, if one considers the set of real numbers and the operation of addition, this assembly turns out to be a group. The elements are combined with each other through the operation, the combination being called the "product." In order to be elements of a group, a set of elements must obey the following rules of combination.

1. The product of any two elements must be an element of the group.
2. The associative law $A(BC) = (AB)C$ must be obeyed.*
3. The group must contain an identity element such that if one operates with a given element of the group on the identity, the same element is generated.
4. Every element must have an inverse, such that if one operates on a given element with its inverse, the identity is generated.

To show that the set of real numbers constitutes a group under the operation of addition, we need only show that the real numbers combine according to the above four rules. Certainly, the sum of two real numbers is another real number (rule 1). The sum of any three real numbers is the same whether we add the sum of the first two to the third or the first to the sum of the

---

*The commutative law $AB = BA$ may or may not be obeyed.

second and third (rule 2). Each negative number is the inverse of each positive number and vice versa (rule 4), provided that the zero is taken as the identity element (rule 3).

The value of group theory in molecular spectroscopy, and in physics in general is not its application to number theory. The quantities of interest in group theory are the *symmetry elements* of an object. It is intuitively obvious that a snowflake is more symmetrical than, say the branch of a tree. Consider, now, the equilateral triangle of Figure 1.4. Imagine a mirror to lie along the

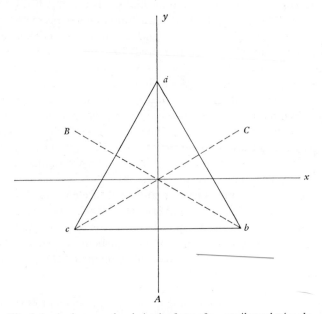

Fig. 1.4 A planar molecule in the form of an equilateral triangle.

y axis, perpendicular to the plane of the paper. If the triangle is now re-flected in this mirror, point a will go into point a, point b will go into point c. and point c will go into point b. Since the three points are identical, the triangle will be a configuration after the reflection, which is indistinguishable from its original configuration. The reflection plane is a symmetry element and the reflection itself, a symmetry operation. Examination of the figure shows it to possess the following elements of symmetry:

1. The identity operation E, which consists of doing nothing to the triangle.
2. Element A-a reflection plane lying in the yz plane.
3. Element B-a reflection plane passing through point b and perpendicular to the line joining a and c.

4. Element C-a reflection plane passing through point c and perpendicular to the line joining a and b.

5. Element D- the clockwise rotation of $120°$ around the z axis.

6. Element F- the counterclockwise rotation of $120°$ around the z axis.

To each of these symmetry elements corresponds a symmetry operation. These symmetry operations can be shown to be the elements of a group, the group operation being simply, the application of a symmetry operation. If the points at the corners of the triangle are given dummy labels, a, b, and c, applying any two symmetry operations in succession is equivalent to some one symmetry operation. The combination characteristics of the group are summarized in the multiplication table.

	E	A	B	C	D	F
E	E	A	B	C	D	F
A	A	E	D	F	B	C
B	B	F	E	D	C	A
C	C	D	F	E	A	B
D	D	C	A	B	F	E
F	F	B	C	A	E	D

$$(1.4.1)$$

The meaning of the table is the following. If an operation along the top is applied to the Figure, then an operation along the left side, the result is the operation at the intersection of the row and column of the operations used. The commutative law does not hold for this group. From (1.4.1), $AB = D$, but $BA = F$, so $AB \neq BA$.

The point operations, or the symmetry elements of a finite figure, do not constitute all of the possible types of symmetry operations. Infinite objects, like an infinite crystal lattice, also have translational elements of symmetry. A fully extended polyethylene chain, shown in Figure 1.5 is an

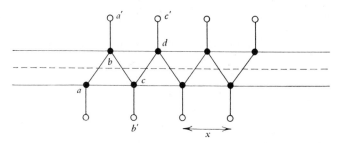

Fig. 1.5 The polyethylene chain. Solid circles: carbon atom; open circles: occluded pair of hydrogen atoms.

example of a one-dimensional crystal. A translation of the entire chain sends point a into c, b into d, the pair of hydrogen atoms a' into the pair c', etc. Polyethylene also has point symmetry operations. A plane of reflection containing carbon atom c and the two hydrogen atoms b' is a symmetry element of the chain, as is any such plane containing a hydrogen atom and two carbon atoms.[4]

In addition to pure translations and pure point operations, an infinite chain or lattice may have mixed operations. These are called glide reflections or screw rotations. In Figure 1–5, the distance x is the unit translation. Now, suppose that the entire chain is translated a distance $x/2$ and rotated 180° around the translation axis. The new configuration will be indistinguishable from the old, so that the screw rotation is a symmetry element. Alternatively, the chain may be translated a distance $x/2$ and reflected in a plane perpendicular to the plane of the paper and containing the translation axis. The new configuration is indistinguishable from the old, so that the glide plane is a symmetry element. The entire assemblage of symmetry operations of polyethylene is called a "line group" or one-dimensional space group.[4] The pure translations constitute a group by themselves. This is a "subgroup" of the full line group. The same principles apply to crystals, except that now the translations can occur in three dimensions.

The next concept to consider is that of a "class." Any element X of a group has an inverse, X^{-1}. Suppose that two elements of a group P and Q, have the property $X^{-1}(P \text{ or } Q)X = P \text{ or } Q$, where X is any element of the group. P and Q are said to belong to the same "class." For the group of equation 1.4.1

$$
\begin{array}{ll}
EDE = D & EFE = F \\
ADA = F & AFA = D \\
BDB = F & BFB = D \\
CDC = F & CFC = D \\
FDD = D & FFD = F \\
DDF = D & DFF = F
\end{array}
\qquad (1.4.2)
$$

Continuing with the remaining elements of the group we find that, the three classes are (D, F), (A, B, C) and (E).

To progress further, we must define matrix multiplication. A matrix is an array of numbers

$$
A = \begin{vmatrix} a_{11} & a_{12} & \cdots & a_{n1} \\ a_{21} & a_{22} & \cdots & a_{2n} \\ a_{n1} & a_{n2} & \cdots & a_{nn} \end{vmatrix}
\qquad (1.4.3)
$$

We may define a second matrix

$$
B = \begin{vmatrix} b_{11} & b_{12} & \cdots & b_{1n} \\ b_{21} & b_{22} & \cdots & b_{2n} \\ b_{n1} & n_{n2} & \cdots & b_{nn} \end{vmatrix}
\qquad (1.4.4)
$$

If the matrices are both square, that is, the number of rows equals the number of columns, and if both matrices are of the same size, n, we can define a new matrix of size n, the product matrix C. The elements of C are

$$c_{ik} = \sum_j a_{ij} b_{jk} \tag{1.4.5}$$

(The reader who is unfamiliar with matrix operations should refer to General References R1 and R6 before going on). The discussion of group theory so far has been couched in terms of abstract elements or in terms of symmetry operations. In order to apply group theory to spectroscopy, it is necessary to find "matrix representations" of the group. These are sets of unitary matrices (matrices whose determinants equal one) which can be set into one-to-one correspondence with the elements of the group. For a group with six elements for example, there will be six matrices in a given representation. The matrices are chosen so that if their matrix products are formed by means of equation 1.4.5, they generate the group multiplication table. For example, in equation 1.4.1 if the matrix assigned to B is multiplied by the matrix assigned to A, the product should be matrix assigned to D. For a given group, an infinite number of representations may be written down. This may be done by inspection, or by using various standard schemes. One such scheme (see below) is to draw the cartesian coordinates of the atoms of a molecule having a given symmetry and to apply the symmetry operations to the cartesian coordinates. If this process is written symbolically in matrix notation, the matrices representing the transformations will be group representations.

In equation 1.4.6 are given three of the infinity of possible matrix representations of the group of equation 1.4.1

	E	A	B
A_1	1	1	1
A_2	1	-1	-1

$$E \quad \begin{pmatrix} 1 & 0 \\ 0 & 1 \end{pmatrix} \quad \begin{pmatrix} -1 & 0 \\ 0 & +1 \end{pmatrix} \quad \begin{pmatrix} \dfrac{1}{2} & -\dfrac{\sqrt{3}}{2} \\ -\dfrac{\sqrt{3}}{2} & -\dfrac{1}{2} \end{pmatrix}$$

	C	D	F
A_1	1	1	1
A_2	-1	1	1

$$E \quad \begin{pmatrix} \dfrac{1}{2} & \dfrac{\sqrt{3}}{2} \\ \dfrac{\sqrt{1}}{2} & -\dfrac{1}{2} \end{pmatrix} \quad \begin{pmatrix} -\dfrac{1}{2} & \dfrac{\sqrt{3}}{2} \\ -\dfrac{\sqrt{3}}{2} & -\dfrac{1}{2} \end{pmatrix} \quad \begin{pmatrix} -\dfrac{1}{2} & -\dfrac{\sqrt{3}}{2} \\ \dfrac{\sqrt{3}}{2} & -\dfrac{1}{2} \end{pmatrix} \tag{1.4.6}$$

It should be noted that the symbol E is used with two different meanings in equation 1.4.6. Across the top, E is the identity element. At the bottom, it is standard spectroscopic symbol for a 2×2 representation. The reader is urged to verify that these matrices indeed combine according to equation 1.4.1. A few words as to how 1.4.6 was generated might be of interest. Every group will have one representation in which all of the matrices are (1). This representation is given the symbol A_1 by spectroscopists. The representation E contains the transformation matrices of the components of a vector which is subjected to the transformations of the symmetry group corresponding to (1.4.1). Representation A_2 was written down by inspection, taking into account the properties of irreducible representations, which will be discussed shortly. These particular examples were selected from the infinity of possible matrix representations because they have the property of being "irreducible."

To explain irreducibility, we need the concepts of a diagonal matrix and an inverse matrix. All of the elements of a diagonal matrix are zero, except elements along the diagonal. That is,

$$a_{ij} = 0, \qquad i \neq j \tag{1.4.7}$$

If

$$\begin{aligned} a_{ij} &= 0, & i \neq j \\ a_{ij} &= 1, & i = j \end{aligned} \tag{1.4.8}$$

the matrix is said to be a unit matrix. A matrix A^{-1} is said to be the inverse of a matrix A if their product, defined by (1.4.5) equals the unit matrix, defined by (1.4.8).

Suppose now, that we have a representation of a group, consisting of a set of matrices, X. It is not hard to show that if we have an arbitrary matrix A and its inverse A^{-1}, that the set of matrices $A^{-1}XA$ is also a representation of the group. It is assumed, of course, that X and A have the same dimensionality and the product (XA) is first formed by using (1.4.5), so $A^{-1}XA = A^{-1}(XA)$. It is conceivable that some matrix A exists such that all of the matrices $A^{-1}XA$ take the form of square blocks along the diagonal with zeros elsewhere. All of the matrices in the representation must of course, have blocks of the same dimensionality in the same position. If no such matrix A exists, the representation is said to be "irreducible." The representations A_1, A_2, and E of equation 1.4.6 are all irreducible.

There are some useful theorems on irreducible representations.

1. The number of irreducible representations equals the number of classes of a group.

2. The sum of squares of the dimensions of the irreducible representations equals the number of elements in the group.

Looking at (1.4.6), we see that there are three irreducible representations, corresponding to the three classes. The sum of squares of dimensions of the irreducible representations is $1^2 + 1^2 + 2^2 = 6$, the number of elements in the group.

For most (not all) purposes, all that is needed is the characters of the irreducible representation, the sum of the diagonal elements of the matrices. All elements belonging to a given class of the group have the same characters, so only one entry need be made per class. The character table for the group of (1.4.1) is

	E	(D, F)	(A, B, C)
A_1	1	1	1
A_2	1	1	-1
E	2	-1	0

$$(1.4.9)$$

We have now reached the point where we can start to tie group theory into vibrational spectroscopy. The connection can be made either classically or through quantum mechanics. Suppose we have a molecule orientated with respect to a given coordinate system and know all of its wave functions, referred to the same coordinate system. Suppose now, that we effectively change the coordinate system by applying a symmetry operation to the molecule. The wave functions must also change. Suppose that j wave functions have the same energy, E_j. It is easy to show that a transformed wave function, u'_k must be a linear combination of the j untransformed wave functions.

$$u'_k = \sum_j a_{kj} u_j \qquad (1.4.10)$$

Since there are j transformed functions, u'_k, it is clear that the transformation will generate a matrix, (a_{kj}). If we apply all of the symmetry operations, we will get a set of matrices (a_{kj}), one for each element of the group. It turns out that a set of matrices generated in this fashion is an irreducible representation of the group describing the molecule. Each energy level can thus be characterized by the group irreducible representation generated by its wave functions. Furthermore, even in cases where Schrödinger's equation is too difficult to solve, group theory determines the symmetry properties of the (unknown) wave functions.

In the classical case, it may be shown that each normal vibration of a molecule has a characteristic symmetry described by an irreducible representation of the group describing the molecule.

We have now reached the point where spectroscopic activity can be discussed. The cartesian coordinates x, y and z and their products, x^2, y^2, xy, etc. also transform in a characteristics fashion under the symmetry operations of a group. Consider a matrix element

$$\int u_j^* x u_i \, dv \qquad (1.4.11)$$

In general, the initial state u_i will be the ground state of the molecule. This will be totally symmetric under the operations of the group. In order for the matrix element (1.4.11) to be unchanged under the symmetry operations, u_j^* and x must belong to the same irreducible representation. Otherwise, the matrix element will change sign under the symmetry operation. If neither x, y or z belongs to the same irreducible representation as u_j^*, no matrix element of the form (1.4.11) can differ from zero. The molecular transition associated with (1.4.11) cannot give rise to the absorption of light. Now consider matrix elements of the form

$$\int u_b^* x u_j \, dv \int u_j^* y u_a \, dv. \qquad (1.4.12)$$

As before, u_a is presumed to be totally symmetric. When the matrix element is changed by a symmetry transformation u_j makes no contribution, since it acts through the totally symmetric product $u_j^* u_j$. For (1.4.12) to be different from zero, u_b^* must belong to the same irreducible representation as the product xy. The condition that the Raman transition ab take place is thus that the final wave function u_b^* belong to the same irreducible representation as the product of two Cartesian coordinates.

At this point, the reader will appreciate the beauty of group theory. In general, the mathematical form of the wave functions u_a and u_b will be unknown. Even when they are known, calculation of the matrix elements is a lengthy undertaking. Application of group theory enables one to pick out the matrix elements equal to zero with no calculation at all. This tells right away what optical transitions are possible. If a calculation of intensity of optical absoprtion or Raman scattering is to be undertaken, one is saved the labor of calculating the zero integrals, only to find that they are zero.

A general rule may be derived for systems like the benzene molecule, which have a center of symmetry. The Cartesian coordinates are sent into their negatives by inversion through a center of symmetry, while their products are left unchanged. Therefore in this case, if elements of the form of (1.4.11) are not zero, for a given u_b^*, those of the form of (1.4.12) must be. We thus have the rule that in centrosymmetric systems, a transition giving rise to Raman scattering cannot give rise to optical absorption and vice versa.

The Raman and infrared activities of different symmetry species for the point groups are tabulated in several places[R3, R5] and do not have to be calculated afresh each time. The tabulation is less complete for the space groups describing polymer chains, planar lattices and crystal lattices. Fortunately, the problem of treating space groups can be reduced to a problem in point group theory,[R7] for most cases of interest. This is because it turns out that the matrices of irreducible representations of a space group can always be written as the product of one of a small set of standard matrices ("coset representatives") and diagonal matrices containing only elements of the irreducible representation of the translation group. (The latter are all one-dimensional). The Cartesian coordinates and their products are all completely symmetric under a pure translation. Therefore, only those wave functions, u_b^*, of a crystal which are completely symmetric under a pure translation can give rise to optical absorption or Raman scattering. The translation matrices in the corresponding irreducible representations are all unit matrices. If the assemblage of these unit matrices is considered to constitute a group identity element, the coset representatives and this identity constitute a group. These so-called factor-groups will always have the same multiplication table and thus the same character table, as one of the point groups. The spectroscopic activity can thus be read off a point group table.

If a crystal contains n atoms, it will have $3n - 3$ vibrational normal modes. A crystal containing 10^{23} atoms will have about 3×10^{23} normal modes. These will all contribute to the specific heat, but only a very few (ordinarily, less than a hundred) can contribute to optical absorption or to the Raman spectrum.

It was noted above that each wave function of a molecule or each normal vibration, is characterized by a symmetry species of the point group of the molecule. One might wish to know, how many normal vibrations of a given molecule are characterized by each symmetry class of the point group? The answer to this is given by an examination of the $3n$ Cartesian coordinates required to fix the positions of the atoms in the molecule. If these coordinates are written in column matrix form

$$\begin{vmatrix} x_1 \\ y_1 \\ z_1 \\ x_2 \\ y_2 \\ z_2 \\ \vdots \end{vmatrix} \qquad (1.4.13)$$

and this matrix subjected to the symmetry operations of the group, a transformed set of position coordinates is generated. The set of $3n \times 3n$ transformation matrices connecting untransformed and transformed coordinates are found to constitute a representation of the group.

This representation is reducible, in contrast to the representations of (1.4.6). To see the significance of this, we must consider the meaning of reduction of a representation. A reducible representation of the group we have been considering will consist of six matrices, one for each element of the group. There exists some matrix A, such that if we carry out the operations $A^{-1}XA$ on the six elements X, the six transformed matrices of the group will have their elements in square blocks along the diagonals. The square blocks will consist of the irreducible representations of the group. The number of times each irreducible representation occurs along the diagonal of the reduced representation of the Cartesian coordinates can be shown to equal the number of normal vibrations having the symmetry of the irreducible representation.

Fortunately, it is not necessary to have an actual matrix A to transform a reducible representation. All that is necessary is to have the characters of the reducible and the irreducible representations. Then, the number of times the ith irreducible representation occurs in a reducible representation is

$$n_i = \frac{1}{y} \sum_R (X_i(R)X(R)) \qquad (1.4.14)$$

Here y is the number of elements in the group, $X_i(R)$ is the character of the ith irreducible representation for the symmetry operation R and $X(R)$ is the character of the reducible representation (transformation matrix) for the symmetry operation R.

To get $X(R)$, we first note that only nuclei whose position is unshifted by a symmetry operation can contribute to the character $X(R)$ for that operation. As an example, consider Figure 1-6.

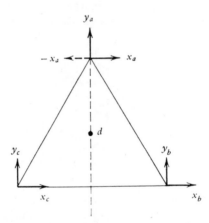

Fig. 1.6 Cartesian coordinates for the molecule of Figure 1.4.

The reflection A sends b into c, c into b and leaves a unshifted. For the unshifted point, a, $z_a \rightarrow z_a$, $y_a \rightarrow y_a$, $x_a \rightarrow -x_a$. The elements along the diagonal of the transformation matrix for the symmetry operation A and associated with point a are, therefore, 1, 1, and -1. (The reader is urged to write down the 9×9 transformation matrix for symmetry operation A, as an exercise.)

Thus $X(A) = (1 + 1 - 1) = +1$. The two rotations D and F leave no nucleus unshifted, so $X(D, F) = 0$. The operation E leaves all three nuclei unshifted. There are three unshifted coordinates per nucleus, so $X(E) = 3 \times 3 = 9$. It is obvious that $X(B) = X(C) = 1$.

A complication would arise were there an atom at point d. The coordinate z_d, is unchanged by the reflections and rotations, but x_d and y_d are transformed into linear combination of x_d and y_d by these operations. Fortunately, the transformation coefficients are the elements of the irreducible representation matrices E of (1.4.6) and the necessary contribution to the character of $X(R)$ can be taken from these. For example, if we consider the rotation D, coordinate z_d would contribute $+1$ to $X(D)$ and the pair x_d, y_d would contribute $(-\frac{1}{2}) + (-\frac{1}{2}) = -1$ to $X(D)$, so that the total contribution of the nucleus at d to $X(D)$ would be zero.

Returning to the equilateral triangle, we have

$$
\begin{array}{ccccccc}
 & E & A & B & C & D & F \\
X(R) & 9 & 1 & 1 & 1 & 0 & 0
\end{array}
\tag{1.4.15}
$$

Using (1.4.9), (1.4.14) and (1.4.15), we quickly find $n_{A_1} = 2$, $n_{A_2} = 1$ and $n_E = 3$. To each of the n_E correspond two normal vibrations of the molecule (these modes are doubly degenerate) so the total number of normal modes is 9, which is correct for a triatomic molecule. Of the nine normal modes, three are rotations and three translations of the molecule as a whole. One translation turns out to have symmetry A, and two E. One rotation turns out to have symmetry A_2 and two E. We are left with one normal vibration of symmetry A_1 and a degenerate pair of symmetry E.

Fortunately, this detailed derivation need not be carried out every time it is desired to classify the normal vibrations of a molecule as to symmetry. Results have been tabulated for the various point groups[R5] and can be read from tables, with little effort.

Application of these methods to the factor group modes of polymer chains[4] or crystals[R7], is only slightly more involved. Only the atoms in a single repeat unit need be considered.

To sum the foregoing, it has been shown that from just a knowledge of the shape of a molecule, the number, symmetry characteristics and spectroscopic activities of a molecule may be calculated. Even these sweeping results do not encompass the entire utility of group theory in molecular spectroscopy. It was indicated in section 1.3, that, if the equations for the normal frequencies of a molecule are set up, a $3n \times 3n$ determinant must be solved for an n atomic molecule, (1.3.3). Even when digital computers are used, this quickly becomes impractical, for large molecules. It is far easier to set the problem up in terms of linear combinations of the original cartesian coordinates. If these are chosen, so that the combinations (symmetry co-ordinates) generate irreducible representations of the molecular symmetry group when subjected to the symmetry operations, the new determinant will consist of blocks along the diagonal and zeros elsewhere. This reduces the problem to the factorization of several small determinants, rather than of one large one.

1.5   INTENSITIES AND SCATTERING CROSS-SECTIONS

Let us consider the following experimental arrangement. A single molecule is bathed in a monochromatic beam of light of flux density I_0 watts cm^{-2}. A detector is set up with a filter to pick up the radiation in a single Raman band only. It is presumed that the detector can be scanned over all solid angles, so that the total scattering into the Raman band can be measured. It is found that the scattered power is given by

$$\Delta I = \sigma I_0 \text{ watts} \qquad (1.5.1)$$

The quantity σ, the total scattering cross section, has the dimensions of cm^2. If N molecules are being illuminated the total scattered power is

$$\overline{\Delta I} = N\sigma I_0 \text{ watts} \qquad (1.5.2)$$

If we now define ρ = number of molecules per cm^3, A = cross-sectional area of the illuminated volume, cm^2, and Δx, the length of the illuminated volume, cm, we have

$$N = \rho A(\Delta x) \qquad (1.5.3)$$

and

$$\overline{\Delta I} = \rho\sigma(\Delta x)AI_0 = k\overline{I}\,\Delta x \qquad (1.5.4)$$

where $\overline{I} = I_0 A$ is the total power in the incident beam. Equation 1.5.4 is simply Beer's law for loss of energy of the incident beam due to Raman scattering into the particular Raman band under consideration and k is the loss coefficient in cm^{-1}.

Other scattering cross-sections can be defined. A Raman band extends over a range of frequencies, so that we can define a differential scattering cross-section, σ' from

$$\sigma = \int \sigma' \, d\nu \qquad (1.5.5)$$

or

$$\sigma' = \frac{d\sigma}{d\nu} \qquad (1.5.6)$$

Likewise, if we wish to take the angular dependence of the scattering into account, we can define

$$\sigma' = \int \sigma'' \, d\Omega \qquad (1.5.7)$$

or

$$\sigma'' = \frac{d\sigma'}{d\Omega}$$

Here $d\Omega$ is an element of solid angle.

The scattering cross-sections of Raman bands are exceptionally difficult to measure. Only a few have been reported in the literature.[5] Some rules of thumb can be laid down regarding the factors governing intensity of scattering. The order of magnitude of σ is 10^{-28} cm². Systems containing large, polarizable atoms or groups, such as Sn, I, C=C, etc., will tend to be strong scatterers. Systems made up largely of ionic bonds, such as silica glasses, tend to be weak scatterers. Equations 1.2.5 and 1.2.9 show that Raman bands of a given frequency will appear on both the high-frequency (anti-Stokes bands) and the low-frequency (Stokes bands) side of the exciting frequency. The ratio of anti-Stokes to Stokes intensity for a given band is

$$\text{anti-Stokes/Stokes} = \left(\frac{\nu_0 + \nu_{ab}}{\nu_0 - \nu_{ab}}\right)^4 e^{-\nu_{ab}/kT} \qquad (1.5.8)$$

where ν_{ab} is the frequency of the Raman shift. The anti-Stokes bands are always weaker than the Stokes band.

A further factor affecting the intensity of the Raman scattering is the resonance Raman effect. If the exciting frequency is near or in one of the electronic bands of the system, the quantity $(\nu_{jb} - \nu)$ in equation 1.2.9 will become small so $|R_{ab}|^2$ and, therefore, the scattering intensity will become large. The ratio of scattering with the exciting frequency in an electronic absorption band of the system to that when the exciting frequency is far from such a band can be as large as 10^6.

1.6   TYPES OF RAMAN SPECTRA

In the foregoing discussion we have made it clear that at least one mechanism by which the polarizability may be modulated is the classical vibration of molecules. The connection with quantum mechanics is made by setting up Schrödinger's equation for the system in terms of the classical normal coordinates. We then expect to see, in the Raman spectrum of a molecule, $3n - 6$ bands associated with molecular vibrations. The symmetry selection rules might forbid some of these to appear, but in the absence of symmetry, the statement is correct.

As might be expected, our treatment was oversimplified. The whole discussion of normal coordinates is based on the assumption of simple harmonic motion. The vibrations of real molecules are not simple harmonic. Fortunately, the anharmonicity is small, so that the treatment is a very good approximation for fundamental modes of vibration. However, overtones or, combination tones of the fundamentals can and do appear in the observed Raman spectrum. Fortunately, these are generally much weaker than the fundamentals.

In pure liquids, without hydrogen bonding or other strong intermolecular forces, the only Raman bands which appear are due to vibrational fundamentals, overtones, or combination tones. In spite of the fact that the theory so far developed was based on the treatment of isolated molecules, it is found to hold very well for liquids.

In gases, the molecules are free to rotate, as well as to vibrate. The rotational energy levels are quantized, so that Raman bands due to pure rotations of the molecule appear near the exciting frequency. In addition, the vibrational bands are accompanied by a rotational fine structure.[R5, R8]

Still another situation exists when one deals with molecular crystals, such as crystalline benzene. As was explained in Section 1.4, the vast majority of the $3n - 3$ internal modes of a crystal cannot have spectroscopic activity. The spectra are determined, in essence, by the contents of a single unit cell. If the unit cell contains m atoms, then $3m - 3$ normal vibrations of the crystal may give rise to infrared absorption or Raman scattering.[6] Of course, if the crystal has elements of symmetry other than pure translations, this number may be reduced still further. In crystals which have relatively weak intermolecular forces, the appearance of the Raman spectrum of the crystal is similar to that of the liquid. However, a group of new, low-frequency bands appears due to vibration of the molecules around their centers of mass, or their restricted translations relative to one another. These "lattice frequencies" are extremely valuable for the study of crystal forces.

In gases, liquids and crystalline solids, we have a special type of situation, where either the symmetry of a single molecule, or that of an entire crystal, governs the features of the spectrum.

In highly polar liquids, in amorphous polymers, or in glasses, this may no longer be the case. In a specimen of glass containing n atoms, it is still true that $3n - 3$ normal vibrations characterize the specimen. However, there is no longer any symmetry restriction forbidding the spectroscopic activity of any of them. If intermolecular forces are very strong, as they are in silica glasses, it is no longer possible even to interpret the spectra in terms of simple molecular groupings. The interpretation of the spectra of such systems is attended by grave difficulties.

A final kind of Raman band which may appear is due to chemical complexing. In a mixture of two components which interact to a moderate degree, the Raman spectrum will generally be a superposition of the Raman spectra of the two components. In addition, new, weak bands associated with complexes may appear. Bands due to the formation of hydrogen bonds may be classed under this heading.

A classification of possible types of Raman bands is given in Table 1.1.

Table 1.1

Gases	Pure rotational
	Rotational-vibrational
	Overtones, combination tones, difference tones
Liquids	Vibrational
	Overtones, combination tones, difference tones
Crystals	Lattice
	Internal-vibrational
	Overtones, combination tones, difference tones
Amorphous	Mixed

1.7 SECOND-ORDER RAMAN EFFECT AND STIMULATED RAMAN EMISSION[7,8]

There are several special types of Raman effects which, although of little use for chemical analysis or structure determination, are of considerable scientific interest. The first of these that will be discussed is the second-order Raman effect. It will be recalled that in equation 1.2.29, the Taylor series expansion for the polarizability was chopped off at the first derivative. Sometimes, particularly in crystals, bands appear which can only be accounted for by taking terms like $(\partial^2\alpha/\partial q_i^2)q_i^2$ into account. Such bands are designated second-order Raman bands. It is important to distinguish between the appearance of overtones in the first-order Raman effect and the second-order Raman effect. In both cases, in contrast to stimulated Raman emission, the polarization is proportional to the exciting field strength. Overtones appear when the system being excited is mechanically anharmonic, second-order Raman bands when it is electrically anharmonic. On occasion, second-order Raman bands may be stronger than first-order bands. The two may generally be distinguished by the greater temperature sensitivity of the intensity of the second-order bands. The second-order Raman spectrum of NaCl is shown in Figure 1.7.

Stimulated Raman emission arises from a mechanism in which the polarization itself is proportional to the third power of the exciting field. It can only be exhibited by illuminating a sample with the megawatts per cm^2 field of a Q-switched ruby laser. An analysis of the effect shows that the polarization at a Stokes frequency is proportional to the Stokes field times the square of the exciting field. As a result, the phenomenon has gain. This is to say that

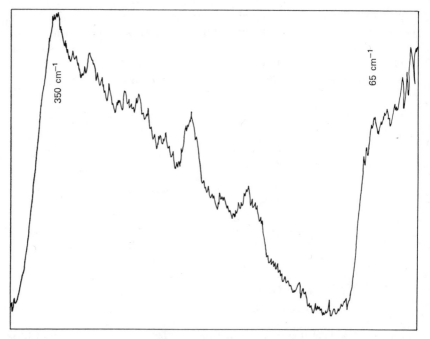

Fig. 1.7 The second-order Raman spectrum of NaCl, excited with a He-Ne laser. The zero of frequency is at the right of the figure. (Courtesy of SPEX Industries.)

as the collimated laser beam proceeds through a long sample, the Stokes field increases exponentially with distance travelled, at the expense of the exciting field.

$$E_{\text{Stokes}} = E_{\text{Stokes}}^0 \exp(aL) \qquad (1.7.1)$$

Since the gain constant a depends on the half-width and intensity of the Raman transition, only one or two of the strongest Raman transitions appear in the stimulated Raman spectrum. For this reason, little information on molecular structure can be gained from the stimulated Raman spectrum. Once sufficient intensity has been built up at the Stokes frequency, various anti-Stokes and higher-order Stokes transitions may take place. The reader is referred to the literature for details.[8]

A potentially more interesting effect is the hyper-Raman effect,[9] in which scattered radiation appears at $2\nu_0$, twice the frequency ν_0 of the exciting radiation and at frequencies displaced from $2\nu_0$ by amounts equal to molecular normal frequencies. Since the selection rules for this hyper-Raman effect are different from those for infrared absorption or Raman scattering.[10] the effect is of great potential interest. Unfortunately, it is very weak and again, must be excited with a Q-switched ruby laser.

2

EXPERIMENTAL METHODS

2.1 INTRODUCTION

In this chapter, perhaps the most important of the book, the experimental techniques used in Raman spectroscopy are described in detail. It is to be expected that the average scientist will buy and use a packaged Raman spectrometer. Nevertheless, for those who, for one reason or another, wish to assemble their own instrument, the methods for doing so are described. Perhaps 95% of the samples which are to be examined, in the typical laboratory, require no special treatment or processing of the resultant spectrum. The remaining 5% do require special consideration. The Raman spectrum of a sample is sometimes accompanied by a powerful, background fluorescence. For one reason or another, it may not be practical to purify the sample. In this event, certain procedures must be used to extract the Raman bands from the background. In other cases, the behavior of a material at high pressure, high or low temperature or other unusual conditions may be of interest. Special sampling accessories are needed for such studies.

The need for adjusting experimental conditions to extract the maximum amount of information and the proper processing of the data is not always appreciated. There is certainly a maximum amount of information that can be derived from a given experimental setup, but it is easy to extract less than this amount. Consideration should always be given to the statistical analysis of the eventual experimental results at the time the experiment is set up. How this is done for Raman spectroscopy will be indicated.

At the present time, a number of good, commercial Raman spectrometers are available. A description of the major features of each is given in Section 2.7.

2.2 SPECTROMETER OPTICS

PRINCIPLES OF PHOTOMETRY

In the simplest terms, the problem of recording an emission spectrum of any kind reduces itself to the problem of transferring radiation from a bright source, the sample, to a detector, a phototube or photographic plate. On the way, the radiation will pass through some fore-optics, a spectrometer and some post-optics. The problem to which we address ourselves is that of getting the maximum flux from sample to detector. It turns out, that, when lasers are used as illuminating sources, this is done most effectively by focusing the laser beam into the sample. The illuminated volume approximates a point source, which greatly simplifies the mathematical expressions. Further simplifications arise from the fact that the sample is on the axis of the optical system, optical elements are perpendicular to this axis and subtended angles are generally small.

Use of the Ebert or Czerny-Turner mount* is quite standard in modern photoelectric Raman spectrometers. The optical layout of the Czerny-Turner mount is shown in Figure 2.1. The centers of the entrance and exit

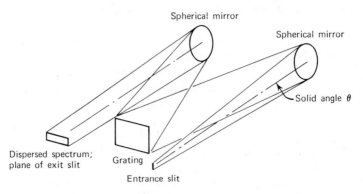

Fig. 2.1 A Czerny-Turner grating spectrometer.

slits lie in the optical plane of the system, at the foci of the concave mirrors. The slits are slightly off-axis from the mirrors. Light coming through the entrance slit is collimated by the first mirror and reflected onto the grating. The dispersed light strikes the second mirror, which brings the spectrum to a focus at the exit slit. The spectrum is scanned by rotating the grating.

It is clear from Figure 2.1 that only rays of light lying wholly within the cone with the entrance slit at its apex and the collimator mirror at its base can get through the spectrometer and out of the exit slit. This cone is called the "cone of aperture" of the spectrometer. The solid angle of this cone is one of the important parameters of the spectrometer. Another is the "dispersion" expressed in angstroms per millimeter at the exit slit. The dispersion is the wavelength interval falling within a millimeter width around the exit slit, when an entrance slit of infinitesimal width is illuminated with white light. Ideally, one would like to have a spectrometer with a large cone of

---

*The two differ only in the use of two mirrors, as shown in Figure 2.1, or of a single mirror.

aperture and high dispersion. Unfortunately, the two features are mutually exclusive, so that some compromise has to be made between angle of aperture and dispersion.

We now turn to a discussion of flux and brightness. Consider a small, bright, rectangular source, which is illuminating an area. We consider only the case where the source and area are perpendicular to the same line through the center of each. Then the flux, in watts, reaching the area is given by

$$F = \frac{B\theta wh}{4\pi} \tag{2.2.1}$$

where $B =$ brightness of source, watts cm^{-2} steradian^{-1}
  $\theta =$ solid angle subtended by the area at the source, steradians
  $w =$ width of source, cm
  $h =$ height of source, cm

In equation 2.2.1, it is assumed that emission is into 4π steradians. If emission is into a hemisphere only, then the factor 4π is replaced by 2π and the value of B adjusted accordingly. The angular variation of the emission is assumed to be averaged out in B. The product θwh is given a special name, the *étendue*[11]

$$E = \theta wh \tag{2.2.2}$$

The *étendue* has the useful property of remaining constant throughout an optical train. To see the significance of this statement, suppose that the illuminated area of Figure 2.2 is a lens, which focuses an image of the source onto a screen. From the definition of a solid angle, the solid angles θ_1 and θ_2, subtended by the lens at source and image, are related by the ratios of the squares of object and image distances.

$$\theta_1/\theta_2 = d_2{}^2/d_1{}^2 \tag{2.2.3}$$

By the same token,

$$w_1/w_2 = d_1/d_2 \tag{2.2.4}$$

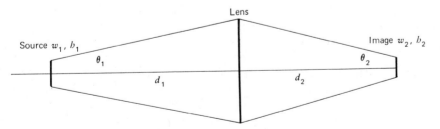

Fig. 2.2 Illustration of the concept of étendue.

and

$$h_1/h_2 = d_1/d_2 \qquad (2.2.5)$$

combining (2.2.3), (2.2.4), and (2.2.5), we find that

$$\theta_1 w_1 h = \theta_2 w_2 h_2 = E \qquad (2.2.6)$$

This is to say, the étendue of each segment of the system is the same. The étendue of a spectrometer is given by the product of the entrance slit height and width on the solid angle of the cone of aperture. In these terms, the entrance slit may be considered as a luminous source, so that the flux which reaches the collimating mirror is

$$F = \frac{BE}{4\pi} \qquad (2.2.7)$$

It goes without saying, that if the spectrometer slit width or height changes, the étendue changes. To illustrate the usefulness of the concept of étendue, let us consider the often-used expedient of back-illuminating a spectrometer. This is to say, a bright source is put just outside the exit slit and the light allowed to pass through the spectrometer, out the entrance slit, through the input optics, to the sample. The path of the light from the input collimating mirror to the sample is shown in Figure 2.3. If one puts a white card outside

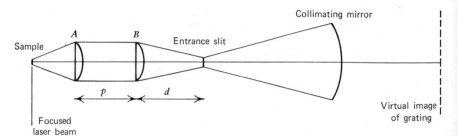

Fig. 2.3   Matching of fore-optics to a spectrometer.

the entrance slit, an illuminated area is visible. Typically, the input optics will have a relatively long focal length lens (B in Figure 2.3) with the entrance slit at its focus. The lens must, of course, have a large enough diameter so that it is bigger than the extended spectrometer cone of aperture, (that is, bigger than the illuminated area just alluded to).

The second lens, A, is of much shorter focal length than lens B. It has the sample at its focus. In a back-lighted configuration, this arrangement produces a demagnified image of the entrance slit at the sample. The illuminated sample volume should be exactly the width and height of this image.

If it is smaller, the spectrometer cone of aperture will not be filled. If it is larger, the image of the illuminated sample volume on the slit will be larger than the slit. Light from the corresponding portions of the sample volume will not get into the spectrometer.

Now, let us assume that sample, input lenses and spectrometer are in the configuration just described but instead of the spectrometer being back-illuminated, the sample is illuminated. Then the étendue of the portion of the system sample-lens A will equal that of the portion sample-image (slit)-lens B, which will equal the étendue of the spectrometer.

It will be noted that nothing was said about the positioning of lens A relative to lens B. Actually, this is not too critical, as long as lens A is larger than the illuminated area coming through lens B when the spectrometer is back-illuminated. There is some advantage to putting lens A at an aperture stop of the system. To locate the aperture stop, it will be recalled (Fig. 2.1) that the grating is *inside* the focus of the collimating mirror. When the spectrometer is back-illuminated, there is a bright, virtual image of the grating behind the collimating mirror. Lens B sees this virtual image as an object, and produces a real image of the grating in space (the slit, in this instance, stops down the aperture of lens B but does not otherwise enter into the optics). Lens A is placed at this image, which is an aperture stop for the system.

At this point, the input optics of a Raman spectrometer using an Ebert or Czerny-Turner mount have been qualitatively described. We now turn to the quantitative considerations necessary to select the lenses.

Equation 1.5.4 gave the total flux scattered into a Raman band by a sample Δx-cm long, illuminated with I watts of exciting radiation. This equation is reproduced here as equation 2.2.8:

$$\Delta I = kI\,\Delta x \text{ watts} \tag{2.2.8}$$

The constant k is the product of the molecular number density, cm^{-3} on the scattering cross-section per molecule, cm^2. If the pickup lens A of Figure 2.3 subtends a solid angle θ at the sample, the flux transferred to the spectrometer is

$$F = \frac{kI\,\Delta x\theta}{4\pi} \tag{2.2.9}$$

Introducing the spectrometer étendue E, we get

$$F = \frac{kI\,\Delta x E}{4\pi w} \tag{2.2.10}$$

where w is the width of the illuminated sample volume. Equation (2.2.10) gives the very interesting result that for a given spectrometer étendue, the flux deliverable to the spectrometer and thus to the detector is inversely proportional to the sample width. This result is valid as long as the image of the sample at the spectrometer entrance slit plane is at least as tall and as wide as the slit. Put another way, when a laser is used as an exciting source, small samples are more efficient than large samples.

The size of the illuminated sample volume is, of course, dependent on the focal length of the lens used to focus the laser beam into the sample.

The diameter of the focal region will be given by the diameter of the diffraction spot,*

$$d = \frac{2.44\lambda f}{D} \text{ cm} \qquad (2.2.11)$$

where λ = wavelength of illuminating light, cm
f = focal length of lens, cm
D = diameter of laser beam, cm

The length of the focal region is generally taken as the Fresnel length, L.[13] This is given by

$$L = \frac{2.44\lambda f^2}{D^2} \text{ cm} \qquad (2.2.12)$$

From (2.2.11) and (2.2.12), the illuminated sample volume is

$$V = \pi \left(\frac{d}{2}\right)^2 L = 2 \times \frac{1.22^3\lambda^3 f^4}{D^4} \text{ cm}^3 \qquad (2.2.13)$$

For a laser beam of given diameter, the length and diameter of the illuminated sample volume are not independent of one another. The ratio of the two is

$$d/L = D/f \qquad (2.2.14)$$

In practice, the width of a spectrometer slit is much smaller than its height. Once a set of input optics has been selected, focusing of the laser beam leads to increased flux throughput only as long as the Fresnel length (after magnification) is taller than the slit height. Once the two are equal, no further gain may be had by using a shorter focal length lens to focus the laser beam into the sample.

MATCHING SPECTROMETER AND INPUT OPTICS: A WORKED EXAMPLE

The foregoing contains all the information necessary to design input optics for a given Ebert spectrometer. A worked example will perhaps give

---

*Actually, the cross-section of the focal region is hyperbolic in shape. The expressions given here are useful, simple approximations[12].

some substance to what will otherwise appear a mass of unconnected equations. The numbers used are taken from the Perkin-Elmer E-1 Monochromator, but the calculation is easily adapted to other instruments. The Perkin-Elmer E-1 Monochromator has an 84×84 mm grating placed 460 mm before a collimating mirror of 580 mm focal length. From the simple mirror formula, the virtual image of the grating will be -2223 mm behind the mirror (Fig. 2.3) Lens B sees this virtual image and produces a real image of the grating at a distance p mm from lens B given, from the simple lens formula, by

$$\frac{1}{p} + \frac{1}{d + 580 + 2223} = \frac{1}{d} \qquad (2.2.15)$$

where $p =$ distance of real image of grating from lens B, mm
$\quad d =$ focal length of lens B, mm

Equation 2.2.15 may be rewritten

$$p = d[1 + (d/2803)] \text{ mm} \qquad (2.2.16)$$

The grating is 84×84 mm, so that the virtual image behind the mirror has an edge given by $(2223/460) \times 84 = 406$ mm. The real image of the grating produced by lens B then has an edge given by (equation 2.2.16)

$$S = 406 \times \frac{d[1 + (d/2803)]}{d + 2803} = 0.145d \text{ mm} \qquad (2.2.17)$$

A second lens, A, is now to be placed at the distance p from lens B. This lens will have a focal length $d' < d$. Therefore, the sample volume will be magnified by the two lenses by an amount

$$M = d/d' \qquad (2.2.18)$$

Equation 2.2.17 gives the *edge* of the real image of the grating. The diagonal of the image is $\sqrt{2}$ times this, or $\sqrt{2} \times 0.145d = 0.21d$. If the pickup lens A is to circumscribe the grating image, it must have a diameter of at least $0.21d$. The f number (ratio of focal length to lens diameter) of lens A is given by

$$f = d'/(0.21d) = 1/(0.21M) \qquad (2.2.19)$$

One would like f to be as small as possible. As a practical matter, $f/1$ is about as fast a lens as is commercially available. Therefore, from equation 2.2.19, $M \simeq 5$ if an $f/1$ lens is used for lens A. If lens A has a diameter of 25 mm and a focal length of 25 mm, then lens B should have a focal length of 125 mm.

The considerations have completely specified the nature of the two entrance optics lenses, A and B. They are in essence, determined by the spectrometer optics and the impracticality of using a lens faster than $f/1$. It only remains to calculate the focal length of the laser focusing lens. The standard slit height for commercial Ebert spectrometers is 10 mm. This means that the Fresnel length should be 2 mm. (After $5\times$ magnification, the Fresnel length will just fill the spectrometer slit height.) Assuming a laser beam diameter of 2 mm and a wavelength of 5×10^{-4} mm, we find from Equation 2.2.12 that $f = 80$ mm for the focal length of the laser beam focusing lens. From Equation 2.2.11, we find the diffraction spot diameter to be 0.0488 mm or 49 microns. The magnified image on the slit has a diameter of about 250 microns. This is the maximum slit width that can be usefully employed with a spectrometer of the optical characteristics just described, if a focused laser beam of wavelength 5000 Å and diameter 2 mm is to be used to excite a Raman spectrum.

The foregoing example shows how to design entrance optics for a given Ebert monochromator. In a commercial Raman spectrometer, these considerations will have been taken into account in designing the spectrometer. The calculation has neglected the effect of the refractive index of a liquid sample. In view of the wide range of refractive indices which is encountered, it is not practical to design the input optics to take this into account. Laser Raman spectrometers are typically fitted with a beam positioner to place the laser beam at the true image of the slit in the sample.

It is worthwhile emphasizing that the above considerations remain valid for any kind of spectroscopy in which a laser is used as an exciting source. The primary consideration of the analysis was to get the maximum flux into the spectrometer system. This involves having the sample quite close to the optics and having the input lens A subtend a large solid angle at the sample. In some instances, it might be desirable to sacrifice flux in order to have the sample farther from the optics. This is easily done by using a longer focal length lens for A and a longer focal length lens to focus the laser beam into the sample.

2.3   THE SIGNAL-TO-NOISE RATIO

Since design and operation of a Raman spectrometer is so intimately bound up with questions of signal detection, this topic will be treated before going further. Suppose that a Raman spectrometer has been set at the frequency corresponding to a Raman transition. Ideally, the recorder should register only the flux of Raman photons at this frequency. Where no Raman band is present, the recorder reading should be zero. In fact, various sources of stray signal are always present.

1. Any phototube at a temperature above absolute zero delivers a thermionic dark current.

2. Since the Rayleigh scattering is so much stronger than the Raman scattering, some light at the exciting frequency is diffused inside the spectrometer. This escapes through the exit slit and produces a photocurrent, no matter what the nominal frequency setting of the spectrometer.

3. Samples often emit fluorescence as well as Raman radiation. Fluorescence is typically broad-band and appears at the same frequencies as Raman bands, as well as at intermediate frequencies.

The phototube dark-current is, of course, independent of the particular frequency interval being viewed. Stray Rayleigh light and fluorescence are only weakly dependent on the frequency at which the spectrometer is set.

When these sources of stray signal are present, the Raman bands appear as peaks on a relatively smooth background (Fig. 2.4). There is a widely

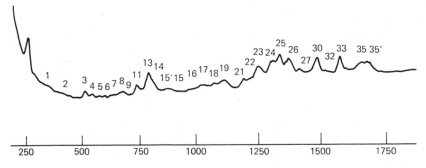

Fig. 2.4 The Raman spectrum of solid salmon testes DNA, showing Raman peaks on a fluorescence background.

held misconception that as long as a background is "smooth," it can be removed electronically by using "zero-offset" on the recorder and greater gain in the amplifier. To see the fallacy in this argument, consider the following experiment. Suppose that a Raman band is being detected with a photomultiplier tube. Whenever a photon activates the cathode, a burst of photoelectrons traverses the tube. Instead of measuring the net photocurrent, the number of individual bursts, or pulses, of the photoelectrons arriving at the anode per second is counted.* In the absence of dark current, this number is proportional to the number of photons arriving at the cathode per second. It is found that there is a statistical fluctuation in the number of pulses counted per second. This fluctuation is built in by nature and nothing can be done about it. If one makes a great many observations of one second

*This approach is the basis of photon counting detection, to be described below.

duration each, an expected, or mean number of bursts, or pulses, arriving per second can be estimated. Let us call this number N_0. Then, if a one-second interval is selected at random, the probability that m pulses were counted during it is

$$P(m) = N_0{}^m \exp{(-N_0)}/m! \qquad (2.3.1)$$

Equation 2.3.1, the formula for the Poisson distribution[R6] governs many natural phenomena, including the rate of emission of α-particles and the occupancy rate of hospital beds.

For fairly large N_0, the Poisson distribution approaches a normal distribution of standard deviation $\sqrt{N_0}$. By convention, if a mean of N_0 pulses per second is being recorded, the "noise" is defined as $\sqrt{N_0}$. From the properties of the normal distribution, a perfectly definite meaning can be assigned to this concept of noise. Again, let us suppose that a great many one-second intervals have been recorded and the number of pulses in each counted. Then approximately 5% of the counted intervals will contain a number of pulses outside the range $N_0 \pm 2\sqrt{N_0}$ and approximately 0.2% of the intervals will contain a number of pulses outside the range $N_0 \pm 3\sqrt{N_0}$.[14]

The foregoing exposition is simply a special case of the general principle that no physical measurement can be made with perfect accuracy. The purpose of statistics is to give a quantitative estimate of the degree of accuracy.

We now wish to define the signal-to-noise ratio of a measurement. This is simply given by

$$S = N_0/\sqrt{N_0} = \sqrt{N_0}/N_0 \qquad (2.3.2)$$

The purpose of design of an experiment is to make this ratio as large as possible.

With this brief introduction, we may now return to the discussion of Raman spectra. Suppose that a Raman spectrometer is scanning a spectrum, spending one second on each frequency interval. Suppose, further, that in parts of the spectrum where no Raman bands are present, the number of pulses due to stray exciting light, phototube dark current and fluorescence has the average value N_0. The recorder trace will be irregular, because of the statistical fluctuation described by equation 2.3.1. When a frequency interval containing a Raman band is reached, the mean number of pulses recorded per second will rise above N_0. The situation is illustrated in Figure 2.5. The fundamental problem is to detect the Raman band above the noise, or irregular background. By now, the reason that "zero-offset" does not help should be obvious. Zero offset will move the trace down toward the zero

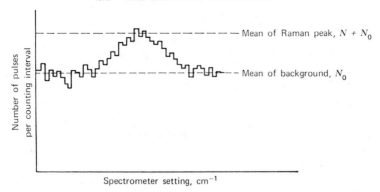

Fig. 2.5   A Raman band in the presence of background noise.

of amplitude, but does nothing towards removing the noise on the background.

The calculation of the signal-to-noise ratio in a Raman band is perfectly straightforward. Since what is desired is to see the Raman band above a smooth background, the signal consists of the N pulses due to Raman radiation only. The noise arises from the N Raman pulses plus the N_0 background pulses. We then redefine the signal-to-noise ratio as

$$S_{\text{Raman}} = N/\sqrt{N + N_0} \tag{2.3.3}$$

Some simplifications of (2.3.3) may be made, in special cases. In the first place, phototube dark current can be reduced to a level which can be neglected, in most cases. (See Section 2.5). The remaining sources of background, the stray exciting light and fluorescence, are both proportional to the intensity of the exciting light, as is the intensity of the Raman emission. Therefore, in a given set of circumstances, the background and Raman intensities are proportional to each other. That is,

$$N_0 = aN \tag{2.3.4}$$

Setting (2.3.4) into (2.3.3) we get

$$S = N/\sqrt{N(a + 1)} \tag{2.3.5}$$

This formula is valid when phototube dark current can be neglected in comparison to other sources of noise. Let us suppose, in addition, that the background is strong, compared to the Raman signal. Then

$$S \simeq N/\sqrt{aN} = \sqrt{N/a} \tag{2.3.6}$$

It will be recalled that N is the number of Raman signal pulses counted per second in a frequency interval of the spectrum. Instead of spending one second in scanning a frequency interval, one might chose to slow down the spectrometer and spend four seconds scanning the same interval. From formula 2.3.6 it is apparent that this would double the signal-to-noise ratio, since four times as many pulses would have been accumulated. On the other hand, if four times as powerful an exciting source were used with a one-second scanning interval, the same doubling of the signal-to-noise ratio would be observed.

Equation 2.3.6 exhibits explicitly the deleterious effect of background radiation on the signal-to-noise ratio. The experimental arrangement should be such as to minimize the background-to-signal ratio, a.

This simple treatment by no means exhausts the power of the statistical approach. Statistical analysis has been used to unravel sources of error in depolarization ratio measurements,[15] to compare photographic and photo-electric recording[16] and to estimate the probability that at least one false Raman band will appear in a Raman spectrum.[14]

2.4 LASERS FOR RAMAN SPECTROSCOPY

The selection of a laser for use as a Raman source depends on several considerations. At the time of writing of this book, argon, krypton, helium-neon, cadmium and ruby continuous-output lasers are on the market and all in use as Raman sources. A list of the principal useful lines of these lasers is given in Table 2.1. No one laser is universally useful for all types of samples. A sample which absorbs strongly in the green region of the spectrum will tend to heat up when exposed to the Argon 4800 Å laser lines. Unfortunately, no photocathode material presently available is very efficient in the 6300–9000 Å region of the spectrum which must be viewed when the He–Ne 6328 Å or the ruby 6943 Å laser lines are used to excite a Raman spectrum. This limits the signal-to-noise ratios attainable in spectra obtained using these sources.

Stability of output and useful operating life are further considerations. The helium-neon laser is by far the best of the available lasers, in this regard. Unfortunately, the output from a helium-neon laser of convenient size does not exceed 80 milliwatts. Nevertheless, the helium-neon laser is standard on commercial Raman spectrometers, because of its reliability.

The ruby laser is useful primarily for deeply-colored samples. The output of 1 watt is attractive, but one is faced with very poor photocathode efficiency in the region of the spectrum beyond 6943 Å.

From the point of view of high power output in a region of the spectrum where efficient photocathodes are available, the argon laser is the Raman source of choice. An added advantage is that krypton can be used in the same laser system, so that the lines at 5682 and 6471 Å are available for colored

Table 2.1

Laser	Wavelength, Å	Typical Power
He–Ne	6328	80 milliwatts
Ruby[a]	6943	1 watt
Argon[a]	4880	0.5–1 watt
	5145	0.5–1 watt
Krypton	5682	0.5 watt
	6471	0.5 watt
Cadmium	4416	0.2 watt

[a] Also available in pulsed mode for ultrarapid spectroscopy. Pulses of 1 to 2 nanosecond duration are typical.

samples. At present, the argon laser is much more temperamental then the helium-neon laser. For the majority of samples, the Raman spectra excited by a typical 500 mw argon line are far superior to those excited by a 80 mw 6328 Å helium-neon line. These superior spectra are paid for by the greater downtime of a typical argon laser. It should be stated that the quality of argon lasers is steadily improving, so that these remarks will not remain valid indefinitely. The question of reliability aside, the argon laser with krypton option is, by far, the best currently available Raman source.

2.5. PHOTOMULTIPLIER TUBES

The recent development of high-quality photomultiplier tubes has been a major factor in the renaissance of Raman spectroscopy. Two characteristics are to be sought in selecting a photomultiplier tube for the detection of weak light signals:

1. High quantum efficiency. The quantum efficiency is the ratio of the number of signal pulses which appears at the anode per second to the number of photons which reaches the photocathode per second. It is a function of wavelength.

2. Low thermionic dark current. In the absence of light, few thermally excited electrons should leave the cathode or dynodes of the photomultiplier tube.

So far as the first requirement is concerned, several photocathode surfaces give quantum efficiencies of 10–20% in the blue and green portions of the spectrum. The situation is worse in the red and near infrared, where quantum efficiencies do not go much above one half of 1%. Nevertheless, the available photocathode materials are entirely adequate for the detection of Raman spectra excited with helium-neon, krypton, or ruby lasers. The most widely used photocathode surfaces are the S-11 and S-20. The S-20 photosurface has a somewhat higher dark noise level than the S-11 photosurface. It has, however, the advantage of being useful over the entire range from 3000 to 8000 Å. The S-1 photosurface is used primarily as a detector when the ruby 6943 Å laser line is used for excitation. Quantum efficiencies for various photosurfaces are illustrated in Figure 2.6.

While there is some variation from tube to tube, several firms now market photomultiplier tubes, which at room temperature, give about five dark pulses per second with S-11 photocathodes and about 100 dark pulses per second with S-20 photocathodes. If the photomultiplier tubes are cooled to $-30°C$, these figures become perhaps one dark pulse every five seconds, and three dark pulses per second, respectively. Only in rare instances does this minute dark signal have to be taken into account.

The S-20 photosurface is the most popular for Raman spectroscopy, because of its wide spectral range of sensitivity. It is a relatively simple matter to cool the photomultiplier tube if ultralow dark noise is desired.

A list of commercially available photomultiplier tubes useful in Raman spectroscopy is given in Table 2.2.

It is desirable to emphasize the relationship between quantum efficiency and dark noise as they affect signal-to-noise ratios. Suppose that one is working under circumstances where the phototube dark current is the major source of background noise. This noise will be independent of the light intensity reaching the photocathode. The overall phototube quantum efficency is defined as the ratio of the number of photoelectron pulses leaving the anode to the number of photons reaching the cathode. The number of photoelectron pulses leaving the anode per second is, then

$$\mathcal{N} = \phi n \qquad (2.5.1)$$

Here ϕ is the quantum efficiency and n the number of photons reaching the cathode per second. If $\mathcal{N}_0$ dark pulses per second are observed, the signal-to-noise ratio is given by

$$n\phi/(\mathcal{N}_0 + n\phi)^{1/2} \qquad (2.5.2)$$

Table 2.2   Some Useful Commercial Photomultiplier Tubes

1.	EMI 6256SA (S-11 surface)
2.	ITT FW130 "Star Tracker" (S-20 surface)
3.	Bendix BX754 "Channeltron" (S-20 surface, 0.04 in. diameter cathode)
4.	RCA-C31000F (GaP surface, sensitive to 9000 Å)

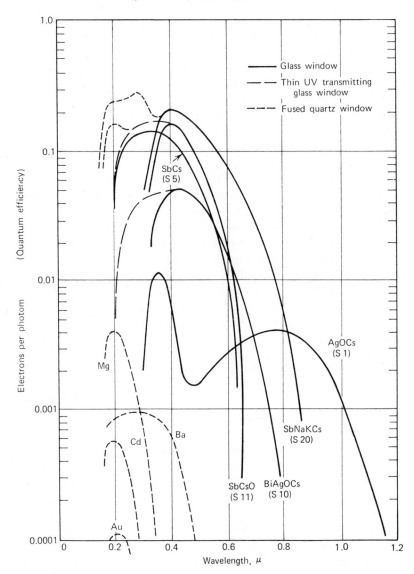

Fig. 2.6   Quantum efficiencies for various photosurfaces as a function of wavelength.

If n is very large, the dark pulse count has little influence on the signal-to-noise ratio. In this case, the photo-cathode surface with the greatest quantum efficiency is the most desirable. However, if n is small compared to $\mathcal{N}_0$, the signal-to-noise ratio depends on the quantity $\phi/\sqrt{\mathcal{N}_0}$. If we are comparing two photosurfaces A and B, where A has twice the quantum efficiency but five times the dark noise of B, photosurface B will clearly give the better signal-to-noise ratios for weak signals.

The best photomultiplier tubes for Raman spectroscopy have small photocathodes, to reduce dark current. It is customary to demagnify the image of the exit slit onto this small photocathode, which may be rectangular in shape. The considerations involved in selecting lenses for the post-optics are identical with those for selecting fore-optics. One feature that needs to be considered is provision for cooling and for positioning the photomultiplier tube. The tube is ordinarily held in a jacket provided with thermoelectric or dry-ice-acetone cooling. It is desirable to have the tube socket mounted in a sleeve in which it may be rotated and moved back and forth. This allows precise positioning of the slit image on the photocathode.

Several cautionary comments should be made on the use of photomultiplier tubes. In the red region of the spectrum the quantum efficiency of the S-20 photosurface decreases almost as rapidly as the dark current, with decreasing temperature.[17] Cooling thus brings only modest increases in the signal-to-noise ratio. Fortunately, photomultiplier tubes with very small photocathodes and thus, very small dark noise are now commercially available.

A photomultiplier tube must never be exposed to bright light when the voltage is on. At worst, this will ruin the tube. At best, there will be a large increase in the dark current. In the latter case, the tube will recover after a days rest in the dark. The voltage on a photomultiplier tube should not be left on when the tube is not in use.

2.6 SIGNAL PROCESSING

The output, of a photomultiplier tube consists of a number of bursts, or pulses of electrons. If the contribution of dark current is neglected, the mean number of pulses per second is proportional to the number of photons arriving at the cathode, per second. The phototube output must be converted to some form, such as displacement of a pen on a chart, which can be read as a plot of intensity versus frequency. Before describing processing methods, something should be said about the characteristics of the pulses making up the output of the photomultiplier tube.[17] Pulses arising from photoelectrons have their origin in electrons leaving the photocathode. These are amplified to varying degrees by the dynodes of the photomultiplier. The pulses thus have a statistical distribution of pulse heights. Their rate of emission per unit time is governed by the Poisson statistics described in Section 2.3. Pulses due to dark current will have a somewhat lower average height than pulses due to photoelectrons. This is because some of the dark current is initiated by electrons thermally emitted from the dynodes. These electrons are amplified by fewer dynode stages than electrons emitted from the photocathode. There is some indication that the rate of emission of dark pulses may, in some circumstances, not follow Poisson statistics. On the average, the total output from the anodes constitutes a current. The individual pulses are, however, easily resolved by modern electronics. There are thus several ways of treating the phototube output. The most common of these are:

1. D-c (direct current) amplification
2. Synchronous (phase-sensitive) detection
3. Noise-voltage detection
4. Pulse counting

In selecting one of these means of processing the phototube output, several considerations come into play. While the processing cannot increase the amount of information contained in the signal, it can certainly decrease it. It is possible, for instance, for an amplifier to contribute its own noise to the output. One or the other processing method might be superior for very strong or for very weak signals. Finally, the cost of electronic equipment varies markedly for the different methods.

The following paragraphs will briefly describe and discuss the advantages and disadvantages of each method.

DIRECT-CURRENT AMPLIFICATION

This is the simplest method of processing the phototube output. A load resistor is put between the phototube anode and ground. The voltage drop produced across the load resistor by the photocurrent is read with a high-impedance electrometer and the output displayed on a recorder. For large signals, this is the preferred method. Most commercial Raman spectrometers have d-c recording as an option. D-c recording suffers from two disadvantages. The first is that it is not sensitive to very weak signals. It would not be indicated if one were examining the Raman spectrum of a gas at low pressures. A more serious disadvantage is that d-c amplifiers suffer from drift, so that the recorder zero changes with time.

SYNCHRONOUS DETECTION

In synchronous detection, the laser beam exciting the spectrum is chopped periodically. A reference signal from the chopper and the output from a load resistor on the phototube are fed to a lock-in amplifier. The lock-in amplifier is sensitive only to signals at the chopper frequency and in phase with it. Therefore, most of the unchopped dark current is rejected by the amplifier.

This method uses a very stable amplifier system. Its major disadvantage is that the laser beam is being viewed only half of the time. This introduces a drop in signal-to-noise ratio of a factor of $\sqrt{2}$, compared to the other methods. A certain amount of noise is also apparently introduced by the amplifier electronics.[18]

NOISE-VOLTAGE DETECTION[19]

This is a relatively new method which is useful for detecting weak signals. It takes advantage of the fact that, if the average value of a signal is viewed as a d-c current and the statistical fluctuations on it as an a-c current, there may, for weak signals, be more time-average power in the a-c component than in the d-c component. A high-pass filter extracts frequencies above 1000 cps from the signal, amplifies them and integrates them.

This method seems to have no advantages over pulse counting, to be described next.

PULSE COUNTING[17, 18]

In this method, the individual photoelectron pulses appearing at the phototube anode are processed. Each pulse is individually amplified. The amplified pulses are next passed through an adjustable discriminator, which rejects pulses below a given height. The pulses passed by the discriminator are used to trigger a pulse generator. The pulse generator produces a standard rectangular pulse at its output, every time the input is triggered. The output pulses may be counted with a scaler, recorded on magnetic tape for further processing, or integrated and displayed on a recorder.

The major drawback of pulse counting is that it fails for very large signals. The equipment can handle, typically, 1 to 10 million pulses per second. A rate of arrival of pulses greater than this jams the electronics. On the other hand, pulse counting offers some major advantages. The data appear in digital form, which is ideal for statistical analysis or for computer processing. With very weak signals, where phototube dark current is the major source or noise, a gain in signal-to-noise ratio may be had by discriminating against the weaker dark-current pulses. Most commercial Raman spectrometers are equipped with pulse-counting electronics.

Several authors[17,18] have made detailed comparisons of these various data processing methods. The consensus of opinion seems to be that pulse-counting gives the best signal-to-noise ratios and higher sensitivity to weak signals of any of the available methods.

Assuming that pulse-counting has been selected as the method of processing the phototube output, one must consider how to convert the output of the pulse counting electronics to a useful form. The most direct method is by use of a digital-to-analog converter. This device produces an output voltage proportional to the number of pulses counted in a preset, fixed time interval.

The output voltages are displayed on a chart recorder. All of the statistical fluctuations in the signal will be displayed in the recorded spectrum. The most common approach is to integrate the output of the pulse-counting electronics with a resistance-capacitance circuit and to display the developed voltage on a chart recorder. This has the effect of smoothing out statistical fluctuations, but may also distort the spectra, if care is not taken. A recently developed method[20] is to record the output of the pulse-counting-electronics on magnetic tape, then to use a computer to carry out a least-squares smooth of the data.

It should be pointed out that these latter two smoothing methods produce a genuine increase in signal-to-noise ratio, not just an apparent increase. This is because they introduce some new information into the system. In essence, these methods require the final spectrum to have a continuous first derivative. Computer processing is somewhat more efficient than resistor-capacitor smoothing, but is also more time consuming.

TIME CONSTANT

If resistance-capacitance integration is being used, the RC or time constant must be matched to the running conditions. Suppose that the spectrometer is set at the peak of a Raman band which gives full-scale recorder deflection at the amplification being used. If the signal is suddenly cut off, the deflection will die as $D = D_0 \exp(-t/T)$, where T is the time constant. The time constant is typically chosen as about one fifth of the time spent in scanning a spectrometer band pass. As an example, if the bandpass is 5 cm^{-1} and the spectrum is being scanned at 2 cm^{-1} per second, the time constant should be 0.5 sec. The effect of the time constant is to smooth shot noise on the trace. If too large a time constant is used, the recorder pen will be unable to respond to the signal, and band shapes will be distorted.

2.7   RAMAN SPECTROMETERS

MONOCHROMATORS

While photographic recording is still sometimes used in the high resolution spectroscopy of gases, the vast majority of the work is done today with a scanning grating monochromator and photoelectric detection. The Raman frequencies are the difference of two large numbers, the frequency of the exciting line and the absolute frequency of the shifted Raman band. This means that the grating drive must be very accurate. As an example, consider a Raman frequency shifted 100 cm^{-1} from an exciting line at 19436 cm^{-1} (5145 Å). The Raman band will appear at 19336 cm^{-1}. If the monochromator drive is accurate to ± 1 cm^{-1}, the absolute frequencies are known to be about 2 parts in 20,000. The uncertainty in the Raman shifted frequency of 100 cm^{-1} is, however, about 4 parts in 100. For this reason, it is necessary to put a calibration spectrum on the Raman spectrum, for accurate work. Neon lamps are useful for this purpose.

It appears that a double monochromator, that is, one monochromator followed by and coupled to a second, is indispensable for serious work in Raman spectroscopy. Even in a carefully designed single monochromator, with careful attention paid to baffling and to low-scatter optics, a large amount of stray exciting radiation is always present in the monochromator. No matter what the nominal frequency setting of the monochromator, this radiation escapes through the exit slit along with the desired Raman radiation. In the second part of the double monochromator, the radiation is redispersed, so that the unwanted exciting radiation is rejected. There is no universally accepted figure of merit for stray light rejection of monochromators. Typically, what is done is to set the slits at some narrow width, say, 10 microns and shine a laser into the monochromator. The transmitted radiation is measured first at the laser frequency, using neutral density filters to reduce the intensity, then at a nominal monochromator setting 10 cm^{-1} from the laser frequency. For a good double monochromator, the ratio of the two measurements should be less than 10^{-9}.

In a double monochromator, both gratings are driven together in tandem. It is essential that they be reproducibly coupled, so as to avoid tracking errors. Commercially available double monochromators are satisfactory in this respect.

The dispersion of the monochromator has already been referred to. Other things being equal, a large linear dispersion at the exit slit is desirable. Large linear dispersion must be paid for by a small cone of aperture of the spectrometer. It is possible to have large linear dispersion and large cone of aperture, but this requires large, expensive diffraction gratings and results in a large, bulky, expensive monochromator. As a result, the commercially available double monochromators all have similar optical characteristics and differ for the most part in such features as the refinement of the mechanical drive. Typically, the linear dispersion between 4000 Å and 8000 Å is about 5 Å per mm. Typical grating dimensions and mirror focal lengths are about 80×80 mm and 600 mm, with 1200 lines/mm the most popular grating spacing. This information for a particular monochromator will be provided by the manufacturer.

Another monochromator characteristic to be considered is throughput. Suppose that a monochromator is being illuminated by a laser beam which has been defocused, so that the cone of aperture of the monochromator is filled. Now suppose that a photomultiplier tube is first put inside the monochromator, just inside the entrance slit, then outside the monochromator, just outside the exit slit. The ratio of the two light intensities measured is the monochromator throughout for the particular laser wavelength. Various losses due to absorption and scattering in the optical elements of the monochromator make the energy leaving the monochromator only a fraction of the energy entering it. A good double monochromator typically has a throughput of about 0.25 between 4000 and 8000 Å.

A few final features of double monochromators for Raman spectroscopy should be noted. As has been stated earlier, these are generally based on the Czerny-Turner mount. Mirror aberrations are minimized for this instrument if curved slits are used. This, however, makes it impossible to image the nearly linear focal region of the laser beam in the sample onto narrow slits. It is preferrable to use straight slits of maximum height of 1 cm and accept the slight loss in resolving power.*

While a continuous drive for the grating is perfectly adequate, a programmed stepping motor is better. This is a motor which sits at a given frequency for a preset time interval, then turns the grating to the next preset frequency. A stepping motor drive is particularly useful when the phototube output is to be processed with pulse-counting electronics.

COMMERCIAL RAMAN SPECTROMETERS

This section presents a brief description of commercial laser Raman spectrometers available on the market as of June 1970. No claim is made to completeness, since the descriptions are based mostly on data provided by the manufacturers. The best instruments are quite similar to one another, using laser sources, double monochromators, low noise photomultiplier detectors and pulse counting detection. There are, of course, differences in refinement of the instruments in such particulars as stability of the wavelength drive, size and accessibility of the sample space, variety of accessories offered, etc. Any of the instruments is probably as easy to use and as reliable as a good quality grating infrared spectrometer. There is no reason why the current crop of laser Raman spectrometers cannot be operated by a skilled technician.

Unfortunately, laser Raman spectrometers follow the general rule that improved performance costs more money. While a 25 to 50 mw helium-neon excitation source is adequate for most applications, difficult samples will prove more tractable when argon excitation is used. The writer strongly recommends that the would-be purchaser of a commercial Raman spectrometer buy it with an argon-krypton source, funds permitting.

Provision for cooling the photomultiplier tube is necessary only when an

---

*It would, of course, be possible to design optics to curve the image of the illuminated volume, but no spectrometer built so far does this.

extremely weak scatterer, such as a dilute gas, is being examined. Otherwise, phototube dark current will be a minor source of noise.

1. Spectra Physics Model 700 (Spectra-Physics, 1250 West Middlefield Road, Mountain View, Calif.). The Spectra Physics Model 700 is a table-top Raman spectrometer designed for simple, routine laboratory use. It is only 40 in. long and 23 in. deep. The source which is provided with the instrument is a 15 mw He-Ne laser, although provision is made for easy coupling to a a Spectra-Physics Model 141 Argon laser. The input optics are equipped with a polarization rotator and an analyser. The monochromator is a 400 mm focal length Ebert double monochromator in a back-to-back additive configuration. The slits are not continuously adjustable. Band passes of 1, 2, 4, and 8 cm^{-1} at 6328 Å are available. In addition, the intermediate slit has a provision for masking the height when microsamples are being examined.

The monochromator is driven by a stepping motor, which can be coupled or uncoupled with a stepping motor driving the recorder.

The signal is detected with an ITT FW-130 photomultipler tube, operated in pulse-counting mode. The electronics automatically switch in an alternate d-c amplifier when the signal becomes sufficiently strong. Provision is made for providing a thermoelectric photomultiplier tube cooler as an option.

Sampling accessories for single crystals, microsamples and powders are available as extra options.

2. Coderg Model PH1 (Coderg, 15 Impasse Barbier-92-Clichy, France). The Coderg Model PH1 is a large, floor model Raman spectrometer. The standard source is an 18 mw He-Ne laser although provision is made for coupling in an argon laser. The double monochromator is of the Ebert type, with a focal length of 600 mm and a speed of $f/8.5$. Curved slits are used. The dispersion is sequential. Detection is with an S-20 surface photomultiplier tube, with optional pulse-counting and d-c readout. The sample space is provided with a kinematic mount. Solid sample, low temperature and high pressure sampling accessories are available.

3. SPEX Ramalog (SPEX Industries, Box 7981, Metuchen, N.J.). The SPEX Ramalog is a large, floor model Raman spectrometer. The standard source provided with the instrument is a 60 mw He-Ne laser. An argon laser is available as an option. The double monochromator is of the tandem Czerny-Turner type, with a 750 mm focal length, a speed of $f/6.8$ and 102×102 mm gratings. The Ramalog monochromator has the unusual (and highly desirable) feature of having an optional provision for introducing photographic recording at the plane of the intermediate slit. Straight slits, with an adjustable mask at the entrance slit, are used. The slits are continuously and independently adjustable with micrometer screws. The instrument has a stepping motor drive and can be provided with readout either in cm^{-1}

or in angstroms. An ITT FW-130 photomultiplier tube is used as a detector. Both pulse counting and d-c detection are provided. Semimicro liquid cells, a capillary cell holder, a crystal powder accessory, and a single crystal goniometer are available as accessories.

4. SPEX Ramalab (SPEX Industries, Metuchen, N.J.). This is a tabletop Raman spectrometer of reasonable cost, specially designed for routine use in the analytical laboratory. It is equipped with a double monochromator, stepping motor drive, S-11 or S-20 phototube response and d-c amplifier processing of the phototube output.

The sample table is kinematically mounted. The instrument comes equipped with two wheels for inserting spike filters or quarter-wave plates into the laser beam, and a polarization scrambler. A particularly attractive feature, available as an option, is a provision for coupling the recorder paper drive to the spectrometer drive. This feature is so designed as to make it possible to record the Raman spectra on standard, preprinted infrared chart paper. The infrared spectrum may, subsequently, be recorded on the same chart. The slits are adjustable to a bandpass of 1, 2, 4 or 8 cm^{-1} at 4416 cm^{-1}. The gratings are 64 $\times$ 64 mm, with 1200 grooves mm^{-1} and an efficiency of 70% at the 5000 Å blaze. The manufacturer claims the gratings to be ghostfree. The monochromator has a speed of $f/7$.

No laser is provided with the Ramalab. The instrument is designed to be coupled easily to a laser of the user's choice.

5. The Perkin-Elmer LR-3 (The Perkin-Elmer Corp., Norwalk, Conn.). The LR-3 is an inexpensive instrument with a tabletop monochromator and rack-mounted electronics and recorder. The monochromator has the Littrow, rather than the Czerny-Turner configuration and is double pass, with a band-stop post monochromator. The radiation is chopped and processed with a lockin amplifier. A 19 mW He-Ne laser is provided with the instrument, although a more powerful laser may be coupled in. A solid sampling accessory and a small-volume cell for liquids are available as accessories.

6. The Jarrell-Ash Model 500 (The Jarrell-Ash Co., Waltham, Mass.). This is a tabletop (39 $\times$ 29 $\times$ 46 in) model, powered by a Coherent Radiation argon laser. Detection is through a pulse-counting system with a linear range of 1.6 $\times$ 10^6 pulses sec^{-1}. Tracking of the gratings of the double monochromator is within 0.3 cm^{-1}, over a range of 3800 cm^{-1}. The sample chamber is unusually large, 15″ $\times$ 9″ $\times$ 22.5″. A variety of sample cells, on pre-aligned kinematic mounts, is available. Optics for depolarization measurements are built in.

7. The Japan Electron Optics Laboratory JRS-Cl. This is a small (5 $\times$ 2 $\times$ 2 ft) tabletop Raman spectrometer powered by a 25 milliwatt helium-neon laser. The instrument has a quoted accuracy of $\pm$3 cm^{-1} from 0 to 4000 cm^{-1}. The readout is direct in wavenumber difference. The

Czerny-Turner double monochromator is unusually fast, $f/4.3$. An optional photomultiplier cooler is available.

2.8 RAMAN SPECTROSCOPY WITH PHOTOGRAPHIC OR LIGHT-AMPLIFIER RECORDING

GENERAL REMARKS

As might be expected, the advantages offered by the use of a double monochromator and photoelectric detection are accompanied by some drawbacks. A typical Raman spectrum covers the frequency interval 0–3500 cm^{-1} The dispersed spectrum is displayed at the plane of the intermediate slit of a double monochromator. A narrow frequency interval of the spectrum passes the slit and enters the second monochromator. The rest of the spectrum is lost. Each resolution element of the spectrum must be examined, in its turn, to record the entire spectrum. Obviously, if a photographic plate or analogous detector were placed at the plane of the intermediate slit, all of the resolution elements of the spectrum could be recorded simultaneously. The SPEX double monochromator may be bought with a camera turret in the first monochromator to do just this. The gain to be realized by using simultaneous, rather than sequential recording has a maximum value of $J^{1/2}$ in signal-to-noise ratio, where J is the number of resolution elements in the spectrum. That is, J is the wavenumber interval covered by the spectrum divided by the spectrometer bandpass. As will be seen below, this maximum gain can seldom be realized. Before considering this question, we will consider detectors for simultaneous recording.

PHOTOGRAPHIC PLATES[21, 22]

Raman spectra are recorded with special photographic plates designed for long exposures to weak light signals. The most popular emulsions are for the red, the Kodak Ia-E emulsion and for the green, the Kodak 1032-O, IIa-O, IIa-J and IIIa-J emulsions. The IIa-O and IIa-J emulsions must be baked for 24 hours at 60°C before use. They have a finer grain than the others. These may be ordered directly from The Eastman Kodak Company, Rochester, N.Y.

Any photographic plate consists of an emulsion of silver bromide crystallites suspended in gelatine, coated onto a glass plate. When light stikes the plate, some of the crystallites are activated. The fraction of crystallites activated is a nonlinear function of the energy per cm^2 which has reached the plate and the wavelength of the light. Once the plate has been exposed, it is placed in a developer bath. Some fraction of the activated crystals, depending on the nature of the developer, the temperature of the developer bath and the development time, is reduced to metallic silver. The developed plate is washed and placed in a fixer bath, which dissolves out any remaining silver bromide. The plate is again washed and dried.

Raman spectra are customarily photographed with an iron or neon emission spectrum, for calibration, above and below the Raman spectrum. The spectrum is read with a microdensitometer, which scans it with an illuminated slit and gives a readout of optical density of the plate versus position on the plate. For maximum information content, the densitometer should have a slit height and width equal to the mechanical slit height and width of the spectrograph used to record the spectrum.

At a given wavelength and for a given energy density of exposure, a typical curve of optical density versus log exposure looks like Figure 2.7a. An unexposed plate always shows a certain degree of fogging, indicated by γ

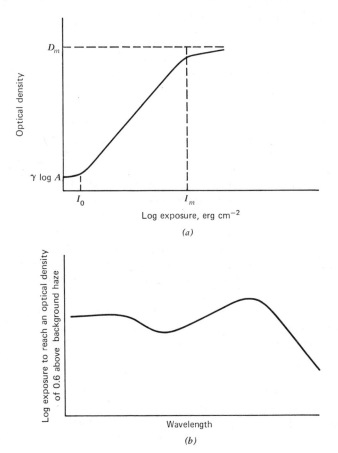

(a)

(b)

Fig. 2.7   Schematic characteristics of photographing plates.

log A in Figure 2.7a. Once the plate is exposed, nothing is discernable above the fogging, until a certain amount of energy, I_0, is provided. The threshold energy, I_0, is called the inertia. Once the inertia has been overcome the optical density is an approximately linear function, with slope γ, of log exposure. Finally, at an exposure I_m, the plate becomes saturated. The behavior of the photographic plate may be conveniently approximated by equation 2.8.1.

$$
\begin{aligned}
D &= \gamma \log A & I < I_0 & \\
&= \gamma \log I + \gamma \log (A/I_0) & I_0 < I < I_m & \quad (2.8.1) \\
&= D_m & I_m < I &
\end{aligned}
$$

where D = optical density
$\quad\quad I$ = exposure, erg cm^{-2}

In terms of this equation, the optical density above fogging is

$$
D = \gamma \log (I/I_0) \qquad I_0 < I < I_m \qquad (2.8.2)
$$

A typical dependence of optical density on wavelength is shown in Figure 2.7b. The manufacture will supply standard curves like those in Figure 2.7 for a given emulsion. These are useful in estimating what exposure time should be used to get a given optical density, once a test plate has been exposed and read.

The signal-to-noise ration of a Raman band on a photographic plate is readily defined. The signal is simply the optical density. The noise is the standard deviation of the fluctuation of optical density on the underlying background. Comparison of the signal-to-noise ratios attainable by photographic recording with those attainable by other methods is extremely difficult, except on a direct experimental basis. Considering just a single resolution element of a spectrum, if one knows the photon flux in the resolution element and the quantum efficiency of the phototube, one may calculate the number of pulses per second appearing at a photomultiplier anode. With the same flux and the optical density versus log exposure curve of the emulsion, one may calculate the optical density to be expected in the plate. Similar calculations may be made for the background underlying the Raman band. Unfortunately there is no a priori way to calculate the noise on this background for a photographic plate.

A few general comments may be made. In scanning a spectrum with a double monochromator, the number of pulses accumulated in a bandpass is given by the number of pulses per second per bandpass divided by the scanning speed in bandpasses per second. This number, divided by the quantum efficiency of the phototube and by the slit area corresponding to the bandpass used, gives the energy density in photons cm^{-2} which would have been delivered in a photographic plate. If this energy density, converted to ergs cm^{-2}, is less than I_0, the plate inertia, the photographic plate would deliver a signal-to-noise ratio of zero. For very weak signals, the photographic plate is a much less efficient detector than a photocathode. If the output of a photomultiplier tube is fed to a scaler, the rate of accumulation of counts, dN/dt, is constant. The rate of increase of optical density with exposure time in a photographic plate, dD/dt, decreases with time and finally drops to zero.[22] These factors, coupled with the inherent graininess of photographic plates, probably more than nullify advantage of simultaneous recording with a photographic plate as detector.

When, then, is the use of photographic recording indicated? In the high resolution Raman spectroscopy of gases, wavenumber positions can be read more accurately from a photographic plate than from a calibrated grating drive. For this application, where signal-to-noise ratios are not so important, photographic recording may be of some advantage. Another application in which photographic recording is useful, is in recording the spectra of transient phenomena. If a chemical reaction takes only a few milliseconds, it might be possible to photograph the Raman spectrum of an intermediate species by using a pulsed laser for excitation. It would be impractical to drive a double monochromator at sufficient speed to record the spectrum.

LIGHT AMPLIFIER TUBES

Although they are still in the development stage, light amplifier tubes offer the possibility of simultaneous recording, without most of the drawbacks of photographic plates. They consist basically, of a photocathode, some amplifying stages and a luminescent screen. Each time a photoelectron leaves the photocathode, a flash appears at the corresponding position on the screen. Bridoux[23] discusses the characteristics and use of several commercial light amplifiers in Raman spectroscopy. As is the case with a photographic plate, the screen of a light amplifier must be read. This is done most effectively by detecting the scintillations in each of the J resolution elements of the spectrum and storing them in a multichannel analyser. No working device described so far does this. Instead, the screen is either photographed, or viewed with a television camera. Either of these approaches involves a loss of information. Nevertheless, it has proved possible to record Raman spectra in fractions of milliseconds, using this kind of detection.[24]

COMPARISON OF SIMULTANEOUS AND SEQUENTIAL RECORDING[16]

Unfortunately, no experimental data are available for the comparison of actual light amplifier tubes with photomultiplier tubes. It is possible, however, to make comparisons between an idealized light amplifier tube and photomultiplier tube. Consider, for example, the following experimental arrangement. A double monochromator, in which the second monochromator rejects stray light, but does not add to the dispersion, has a photomultiplier tube at its exit slit and an optical camera turret at its intermediate slit. The camera turret is fitted with a light amplifier having the same photosurface as the photomultiplier tube. The readout on the light amplifier tube is able to count the scintillations in each resolution element of the spectrum and store the counts in a multichannel analyzer. Now, assume that y minutes are spent in viewing a given Raman spectrum with the light amplifier. Subsequently, the same y minutes are spent to scan the entire spectrum with the double monochromator drive and the photomultiplier tube. If a given resolution element of the spectrum accumulated N counts with photomultiplier recording, it accumulated JN counts with light amplifier recording. Again, J is the number of resolution elements in the spectrum. If the major source of background in the first monochromator is stray Rayleigh light, the number of background counts underlying a Raman band will be proportional to the number of signal counts in that band above background. If the proportionality constant is a, then the signal-to-noise ratio for the Raman band with photomultiplier recording is (Reference 14 and Section 2.3)

$$S_{\mathrm{PM}} = (N/a)^{1/2} \qquad (2.8.3)$$

For the light amplifier, the constant a' giving the ratio of background to signal counts will be larger. This is because the light amplifier is at the intermediate slit and does not benefit from the rejection of stray light by the second monochromator. We find:

$$S_{\mathrm{LA}} = JN/(J/Na')^{1/2} \qquad (2.8.4)$$

and

$$S_{\mathrm{LA}}/S_{\mathrm{PM}} = (Ja/a')^{1/2} \qquad (2.8.5)$$

For typical double monochromators, the ratio (a/a') is about 10^{-4}. Thus, unless J is much larger than 10^4, the added rejection of a double monochromator against stray light overcomes the advantage of simultaneous recording. Part of the advantage of the simultaneous recording can be recovered by fitting the input optics of the Raman spectrograph with a bandstop filter, to reject the Rayleigh light.[25] This should always be done when using either photographic or light amplifier simultaneous recording.

If the major source of background in the light amplifier spectrometer is not stray Rayleigh light, but phototube dark current or fluorescence, then[16]

$$S_{\mathrm{LA}}/S_{\mathrm{PM}} = J^{1/2} \qquad (2.8.6)$$

It must be emphasized that equation 2.8.5 is based on an ideal system with optical readout of the light amplifier and the only source of noise, scatter in the spectrometers. Presently available light amplifiers suffer from scatter in the photocathode optics, which adds noise ("signal-induced noise"). The readout methods used are also less than optimal. It is thus likely that no existing light amplifier Raman spectrometer compares favorably with existing double monochromator-photomultiplier recording. The light amplifier does, at least, have the possibility of being used to deliver a larger signal-to-noise ratio for a given recording time than a photomultiplier tube. It is, at present, the detector of choice for the spectra of transient species.

2.9 DETERMINATION OF RAMAN BAND CHARACTERISTICS

In an ideal Raman experiment, one would end up with a plot of scattering cross-section versus frequency, for two polarizations of the exciting radiation. What is actually recorded on a recorder chart is far from this. For one thing, the spectrometer throughput and the photomultiplier cathode response are highly nonlinear functions of wavelengths. For another, the necessity of using slits of finite width distorts the true shape of the Raman bands. If the only desired information is the position and approximate depolarization ratios of the Raman bands, it is not necessary to correct for these effects. If accurate information on absolute or relative scattering cross-sections is desired, then the Raman spectrometer must be calibrated. The determination of the relative response of a spectrometer system to frequency and corrections for finite slit width are not too difficult to make. The determination of absolute scattering cross-sections is a major undertaking, so that few experimental results have been reported.[5]

For ordinary use, the data that one wishes to derive for a given Raman band, well separated from its neighbors, are the frequency of the peak, the depolarization ratio, the half-width (the width of the band in cm^{-1} at 50% of the peak height) and the peak height or integrated intensity in some arbitrary units.

MEASUREMENT OF RELATIVE WAVELENGTH RESPONSE

The radiation picked up from a small sample by the input optics of a Raman spectrometer is modified by several wavelength-dependent factors by the time it reaches the photocathode surface. Once a Raman spectrometer has been set up, there is no sense in trying to correct for these individually. It is much simpler to determine the response of a Raman spectrometer as an integrated system.

Suppose the spectrometer to be back-lighted. Then, a demagnified real image of the entrance slit will appear in the sample area. A mask may be placed at this image with an opening coincident with the image. If a known flux of energy is caused to pass through this opening, the recorder deflection corresponding to this amount of energy may be read. The known flux of energy is most commonly provided by a standard lamp. Such lamps may be obtained from the National Bureau of Standards or from Electro Optics Associates, Palo Alto, Calif. They are provided with a calibration chart giving the energy per cm² of target area per millimicron of wavelength at a given distance from the lamp. At greater distances, the change in energy density at the target may be calculated from the inverse square law. It is customary to carry out these calculations in terms of wavelength rather than in terms of frequency. This is because the dispersion of a grating is nearly linear in wavelength. For a fixed slit setting, of course, the bandpass of the spectrometer in cm^{-1} will vary with frequency.

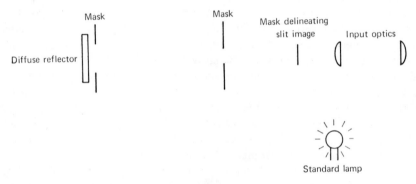

Fig. 2.8   Experimental arrangement for calibrating the wavelength response of a spectrometer system.

An experimental arrangement for doing the calibration is shown in Figure 2.8. A mask is placed in the plane of the entrance slit image, to delineate the sample volume. A block of MgO or other efficient diffuse reflector is placed at some distance from the spectrometer. (This block should be larger than the extended cone of aperture of the spectrometer). The standard lamp is placed so as to illuminate the block at as close to vertical incidence as possible and turned on. The spectrometer and input optics should be shielded so that only light coming through the mask at the sample space can reach any of the input optics. If, now, the spectrum is scanned, the chart will trace out the relative response of the entire system to wavelength. The recorder response can be put at a convenient level by proper positioning of the standard lamp. The calibration should be repeated for a number of slit settings of the spectrometer, to check the dependence of signal on slitwidth. The signal should vary as on the square of the slitwidth.

This system has the advantage that it can be used to calibrate the absolute response of the spectrometer. Suppose, now that masks are placed on the MgO block, just at the edge of the area which if masked leads to a deflection of the recorder. These masks will delineate the area which is contributing energy to the spectrometer. Neglecting a small correction for the fact that different portions of the MgO block are at slightly different distance from the opening in the slit image mask, the flux passing through this opening is

$$T = \frac{BAa}{2\pi r^2} \text{ watts } m\mu^{-1} \tag{2.9.1}$$

Here B is the flux in watts cm^{-2} $m\mu^{-1}$ reaching the MgO block, A is the unmasked area of the block, a is the area of the opening in the slit image mask and r the distance from this opening to the MgO block. The opening in the slit image mask is an effective source for the spectrometer, which emits T/a watts cm^{-2} $m\mu^{-1}$ into the solid angle, θ, accepted by the system. This solid angle may be determined by masking the near face of the pickup lens B and measuring the distance from the slit image mask to the lens mask. The final arrangement gives the response of the entire spectrometer system to a source emitting $T/(a\theta)$ watts cm^{-2} $m\mu^{-1}$ steradian^{-1} into the solid angle accepted by the system.

Once the spectrometer system has been calibrated, it may be used to measure absolute scattering cross-sections of Raman bands. This is best done for liquids by holding the sample in a cell of rectangular cross-section. The results must be corrected for the refractive index of the liquid and losses at the liquid-glass and glass-air interface. Details are given by Skinner and Nilsen.[5]

For most applications, it is not necessary to carry out primary measurements of absolute scattering cross-sections. It is sufficient to refer scattering cross-sections to that of the 992 cm^{-1} benzene Raman band, once the relative response of the spectrometer system has been calibrated. Skinner and Nilsen give a value of $1.05 \pm 0.08 \times 10^{-29}$ cm^2 molecule^{-1} steradian^{-1} $(cm^{-2})^{-1}$ of linewidth at the peak of this band. In their measurement, the laser beam propagated along the z axis with the electric vector along the y axis with observation of the component with electric vector along the y axis. Again, the results must be corrected for the relative refractive index of the sample and of benzene at the wavelengths of the respective Raman bands.

Some typical results for scattering cross-sections are given in Table 2.3.

The reader is cautioned that phototube responses change with time and spectrometer optics drift out of alignment. The calibration for relative response should be checked frequently.

In addition to wavelength responses, the wavelength passed by the spectrometer at a given drive setting may drift from that shown on the drive dial.

Table 2.3. Linewidth, Depolarization Ratio (ρ), and the Relative Peak Value of the Raman Differential Cross Section of a Number of Liquids

Material	Slit	Linewidth, cm^{-1}		$\dfrac{(Nd\sigma_\rho)\text{ Sample}}{(Nd\sigma_\rho)\text{ Benzene}}$		ρ
Methyl iodide	100 μ	5.14	av	0.64	av	
533 cm^{-1}	50 μ	5.34	5.2	0.73	0.7	0.12
Carbon disulfide	100 μ	1.53		2.67		
655 cm^{-1}	50 μ	1.22	1.4	2.65	2.7	0.05
Pyridine	100 μ	2.25		0.60		
991 cm^{-1}	50 μ	2.51	2.4	0.59	0.6	0.02
Iodobenzene	100 μ	1.52		0.83		
999 cm^{-1}	50 μ	1.86	1.7	0.78	0.8	0.05
Bromobenzene	100 μ	1.90		0.64		
1000 cm^{-1}	50 μ	1.98	1.9	0.65	0.6	0.04
Chlorobenzene	100 μ	1.71		0.70		
1002 cm^{-1}	50 μ	1.52	1.6	0.74	0.7	0.02
O-Nitrotoluene	100 μ	16.26		0.30		
1340 cm^{-1}	50 μ	16.40	16.3	0.30	0.3	0.21
Nitrobenzene	100 μ	6.45		1.04		
1345 cm^{-1}	50 μ	6.67	6.6	1.24	1.1	0.15

This should be checked occasionally by scanning the spectrum of a neon lamp or other source with bands of known wavelength.

RAMAN BAND SHAPES AND SLIT FUNCTIONS

Once a spectrometer system has been calibrated for wavelength response, it will not yield true shapes of Raman bands. The difficulty may be seen by considering a spectrometer with slits set at a width of 0.5 mm, with a dispersion of 5 Å per mm at the exit slit. Suppose that helium-neon laser 6328 Å radiation is passed into the spectrometer. Typically, this band has a half-width of several hundredths of an angstrom. An image of the entrance slit 0.5 mm wide will appear at the plane of the exit slit. Now, as the spectrum is scanned, no signal will appear until the spectrometer reads 6325.5 Å (Fig. 2.9a). The signal will then rise to a maximum at a reading of 6328 Å and drop to zero again at 6330.5 Å. The result is that a spectral line with half-width of perhaps 0.01 Å appears to have a half-width of 2.5 Å. Clearly, the apparent half-width will depend on the slit dimensions and the dispersion of the spectrometer. The triangular function in Figure 2.9b is called the slit function. Actual slit functions are slightly rounded by diffraction effects. When a Raman band whose half-width is large compared to the half-width of the slit function being scanned, each infinitesimal frequency element is

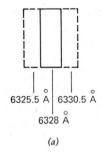

$$6325.5 \text{ Å} \quad | \quad 6330.5 \text{ Å}$$
$$6328 \text{ Å}$$

(a)

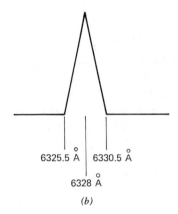

$$6325.5 \text{ Å} \quad | \quad 6330.5 \text{ Å}$$
$$6328 \text{ Å}$$

(b)

Fig. 2.9   Slit function of a spectrometer.

similarly affected by the slit function. The result is, that if the true contour of a Raman band is $F(w)$, and the slit function at a nominal spectrometer setting of w_0 is $S(w - w_0)$, the apparent intensity of the Raman band at w_0 is*

$$I(w_0) = \int_0^\infty F(w)S(w - w_0) \, dw \qquad (2.9.2)$$

Once the slit function has been measured or calculated as described above, and once $I(w)$ has been recorded, $F(w)$ can be retrieved from (2.9.2) by use of Fourier integrals.

As a practical matter, a good estimate of the half-width of a Raman band can be gotten by measuring the apparent half-width at a number of slit settings. A plot of apparent half-width versus slit-width is generally linear

*The width of the exciting line is neglected here. It is almost always much narrower than Raman bands and much narrower than the spectrometer bandpass. The major exception is in the Raman spectra of light gases at low pressure.

and may be extrapolated to zero slit-width for an estimate of the true half-width.[26]

It should be apparent, from the discussion of this and the preceding subsection, that measurements of scattering cross-section or relative intensity may be seriously in error unless the slit function is taken into account. For Raman bands which are broad compared with the spectrometer bandpass being used, this is of no consequence. However, the measured contours of narrow bands must be corrected by use of equation 2.9.2.

The integrated intensity of a Raman band seems to be quite insensitive slit-width.[R2] As with the half-width, it may be measured at a number of slit-widths and a plot of apparent integrated intensities versus slit-width extrapolated to zero slit-width.

THE SHAPE OF RAMAN BANDS

Numerous formulas have been proposed to match the shape of Raman bands. Such attempts really have meaning only when a Raman band is well isolated from its neighbors. It is not uncommon for Raman band peaks to be separated by less than the half-width of the individual bands. In this event, no amount of instrument resolution will separate the bands. The most commonly used functions for describing band shapes are the Gaussian,

$$F(w) = F_0 \exp \frac{(w - w_0)^2}{a} \tag{2.9.3}$$

and the Lorentzian

$$F(w) = \frac{b}{c + (w - w_0)^2} \tag{2.9.4}$$

These are both symmetrical around the frequency, w_0, of the band peak. Each function contains two constants in addition to the peak frequency, w_0, so that a measurement of the peak intensity and the half-width fixes both the integrated intensity and the band shape. The peak heights are $F_{max} = F_0$ in (2.9.3) and $F_{max} = b/c$ in (2.9.4). The half-widths are calculated by setting $F_{1/2} = \frac{1}{2} F_{max}$ and solving the resultant quadratic equations for $\Delta \nu_{1/2}$. Actual Raman bands fit these functions reasonably well near the peak, but the fit may be less good near the wings of the band. Expressions with more disposable constants can be made to give a better fit, but are harder to use. Formulas like (2.9.3) and (2.9.4) are useful when they do give a good fit to true band shapes, because they can be inserted into (2.9.2) with experimentally determined slit functions. A standard table of $I(w)$ can be calculated for a given Raman spectrometer. Once this has been done, true peak heights and half-widths can be read off immediately from apparent ones. These functions are also useful in estimating integrated intensities.[R2] Measurement of the wings of a Raman band is difficult. If there is reason to believe that the shape of a band is Gaussian or Lorentzian the integrated intensity can be calculated from (2.9.3) or (2.9.4) more accurately than it can be measured.

Unless the Raman spectrum is very simple, few Raman bands will be found to be well isolated from their neighbors. The problem then arises of separating out the components of a complex band. If only two unresolved bands are to be separated, this may be done graphically. The procedure is shown in Figure 2.10. The bands are assumed to be symmetrical on a wave number scale around their peaks. The wing of one band is reflected around its peak and plotted under the other band. The ordinates of the reflected wing are subtracted from those of the second band (Fig. 2.10a). The remainder of the second band contour is filled out by reflecting its wing. The contour of the first band is recovered by subtracting that of the resolved band from the overall contour. The consistency of the procedure may be checked by noting whether or not the dotted and dashed portion of (Fig. 2.10b)

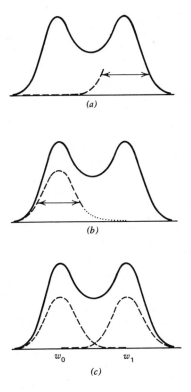

(a)

(b)

w_0 w_1

(c)

Fig. 2.10 Graphical separation of two overlapping Raman bands.

match smoothly, and by repeating the process, starting with the other band.

More complicated Raman bands must be resolved with some such device as the DuPoint curve plotter. This device enables Gaussian or Lorentzian curves of an arbitrary peak frequency, half-width and peak height to be generated, added and exhibited on an oscilloscope screen. The constants are adjusted until the simulated Raman band matches the experimental one. It should be remembered that, since experimental data have finite accuracy, this procedure is, to some extent, aribtrary. A weak band buried under several strong bands could probably not be resolved. Nevertheless, good estimates of band characteristics can often be made by resolving the components. If the components of the band arise from different chemical constituents, the procedure may be checked by varying the compositions of the mixture.

DEPOLARIZATION RATIOS

The measurement of depolarization ratios is subject to all of the errors of the measurement of band contours, plus some additional ones. For the most accurate work the contours of the two components of polarization of the Raman band should be corrected for the effect of finite slit width and spectrometer response. Modern Raman spectrometers are fitted with polarization scramblers at the entrance slit. These are wedges of quartz or calcite which depolarize completely light passing through them. This compensates for any polarizing properties of the spectrometer optics.

Two depolarization ratios may be defined. In both, the sample is illuminated with a beam which is parallel to the spectrometer slit (Fig. 2.11). In a the electric vector of the exciting light is pointed first at the spectrometer slit, then perpendicular to this direction. In b, the electric vector of the exciting radiation is kept parallel to the face of the spectrometer slit. The two components of polarization are separated out with a polaroid placed in the collimated light space of the input optics. This second depolarization ratio is the one most commonly used. The discussion will be based on it.

Before undertaking a depolarization ratio measurement, light from an incandescent lamp or other source of unpolarized light should be passed through the system. In this way, any irregularities in the polarizing properties of the polaroid or inefficiency of the polarization scrambler may be detected. Systematic errors, like those arising from the finite acceptance angle of the input optics, may be detected by measuring the depolarization ratio of a Raman band which is know to be depolarized. The 218 cm^{-1} band of carbon tetrachloride is a suitable band for this purpose. The measured depolarization ratio should be 0.75. A further check may be made by

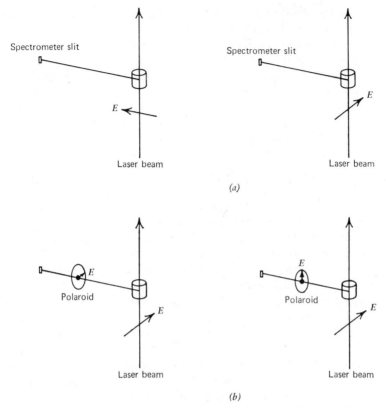

(a)

(b)

Fig. 2.11 The measurement of depolarization ratios of Raman bands.

measuring the depolarization ratio of the 459 cm^{-1} band of CCl_4. This should be 0.01 ± 0.001.

In theory, the depolarization ratio is defined in terms of the integrated intensities of the two components of polarization of the Raman band. For ordinary use, the ratio of the peak heights is adequate. The reader is cautioned not to use the result of a single measurement of the depolarization ratio. A measurement of the depolarization ratio should be repeated a number of times and a mean and standard deviation calculated. This is particularly important when there is a question of whether or not a given Raman band is depolarized ($\rho = 0.75$) or weakly polarized (say, $\rho = 0.73$). If a standard deviation and mean are available, a statistical test, Student's "t" test, may be applied to say whether or not the data are consistent with a mean of 0.75.[R9] Polarization measurements in single crystals, liquids, and gases are straightforward. Crystal powders thoroughly scramble polarizations, so that no significant measurements can be made on these. By immersing the powder in a liquid whose refractive index matches that of the powder, measurements can be made. However, the resultant numbers do not refer to the averaged polarizations of the molecules making up the crystal, but to averaged polarizabilities referred to the crystal axes.

Where the Raman spectra are very weak, there is some advantage to measuring the depolarization ratio of case a, above. Twice the light flux is available in case a as in case b, since there is no loss due to the polaroid. Depolarized bands for this case have $\rho = \frac{6}{7}$. Unfortunately, convergence errors affect case a more than they affect case b.

2.10. SPECIAL METHODS FOR SINGLE CRYSTALS

Single crystals require some separate considerations both from the theoretical and experimental viewpoints. Crystals differ from liquids and gases in that there are reference directions fixed inside the sample. Careful attention must be paid to the exact orientation of both the direction of incidence of the exciting radiation and the direction of observation of the Raman radiation. Crystals of symmetry lower than cubic are birefringent. It is essential, for such a crystal, that the direction of the exciting radiation inside the crystal lie along a principal optical plane. Otherwise, the plane-polarized incident radiation may become elliptically polarized, so the beam may be deviated inside the crystal, even though it is incident perpendicular to a crystal face. By the same token, observation must be along a principal optical direction. At that, the finite angle of observation of the pickup lens will cause difficulties. The reader who is contemplating work with single crystals is urged to review the optical properties of birefringent crystals.[27]

Crystals of orthorombic or higher symmetry have the position of the principal optical directions fixed by symmetry. These will be independent of wavelength. For crystal of monoclinic or triclinic symmetry, the principal optical directions may be wavelength dependent. These questions must all be resolved before an experiment is undertaken.

Studies of single crystals have the advantage that they give a direct value for the polarizability derivatives in a given mode of vibration. The components of the incident electric field and of the polarization (see equation 1.2.7) are all known in such an experiment. Therefore, by using a suitable experimental arrangement, the components of the polarizability tensor may be read off. Depending on the crystal class, some of the elements of the polarizability tensor may be zero for a given symmetry species of the vibration. Loudon[28] and Ovander[29] give tables of the polarizability tensor for the various crystal classes.

Consider, for example, the A species of crystal class C_2. The relation between the electric field and the polarization is[29]

$$\begin{vmatrix} P_x \\ P_y \\ P_z \end{vmatrix} = \begin{vmatrix} a & b & 0 \\ b & c & 0 \\ 0 & 0 & d \end{vmatrix} \begin{vmatrix} E_x \\ E_y \\ E_z \end{vmatrix} \qquad (2.10.1)$$

The form of the polarizability tensor in equation 2.10.1 is fixed by symmetry. Although there are five nonzero tensor elements, only four are distinct. These are labeled a, b, c, and d. Essentially, this means that four experiments are required to evaluate the tensor for this case.

Suppose that the incident radiation is coming along the z axis, with the electric vector along the x axis and observation along the y axis (Fig. 2.12a). Since $E_y = E_z = 0$, we find from equation 2.10.1 that $P_x = aE_x$, $P_y = bE_x$ and $P_z = 0$. Only P_x can be measured, since P_y lies along the direction of observation. Now, suppose the incident radiation to come along the x axis with the electric vector along the z axis and the direction of observation along the y axis (Fig. 2.12b). We find $P_x = 0$, $P_y = 0$, $P_z = dE_z$.

Let us consider the case where the incident radiation is coming along the z axis and both the electric vector and direction of observation are along the y axis. We find $P_x = bE_y$, $P_y = cE_y$, $P_z = 0$. Again, P_y cannot be observed in this configuration (Fig. 2.12c). Lastly, let the incident radiation lie along the z axis, the electric vector along the y axis and the direction of observation along the x axis (Fig. 2.12d). As in the last case, $P_x = bE_y$, $P_y = cE_y$, and $P_z = 0$. Now, P_x cannot be observed.

Looking over the preceeding paragraphs we see that for the four nonzero components of the polarizability of a normal vibration of species A, for crystal class C_2, can be determined in four experiments.

It must be emphasized that the axes which we have been taking are the laboratory coordinate axes, with the crystal remaining in a fixed position during the four experiments. The tensor (matrix) in equation 2.10.1 has the form given only for a specified orientation of the crystal axes relative to the laboratory axes. One must be sure that the crystal is in the orientation specified in the paper from which the tensors are taken. Of course, if one takes care to keep one's direction straight, the crystal can be shifted, rather than the direction of observation.

Current literature on the Raman spectra of crystals uses a notation due to Porto.[30] The symbol $i(yz)j$ means that the incident light is coming along the i axis and the scattered light is being observed along the j axis. The incident light has its electric vector along y and the z component of the scattered light is being observed.

It should be remembered that the polarizabilities determined in Raman experiments in crystals are crystal polarizabilities and not molecular polarizabilities. Even if the spectrum of molecule A in a solid solution of molecule B is being observed, the measured polarizabilities are those of A in the crystal field of B.

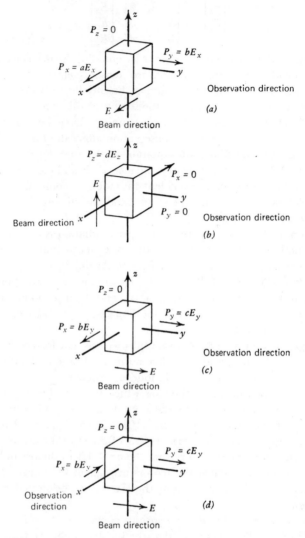

Fig. 2.12 Measurement of the elements of the polarizability tensor of species A of crystal class C_2.

2.11 GASES

Recording the Raman spectra of gases with modern laser Raman equipment presents no difficulty. Indeed, since gases rarely fluoresce and, if filtered, do not have exciting light scattered from dust particles, phototube dark noise is often the major source of noise. The low particle-number-density presents no problem, provided that one is satisfied with 1 cm^{-1} resolution. If one wishes to resolve the vibrational-rotational structure of the Raman bands of light molecules at low pressures, then the experiments become difficult. The most effective expedient has been found to be placing the Raman gas cell with Brewster's-angle windows inside the laser cavity and fitting the laser with totally reflecting windows.[20] The laser beam is focused inside the cell and recollimated with a pair of lenses on either side of the gas cell. Such an approach is equivalent to building a special Raman spectrometer, or at least a special laser Raman source for the study of gases.

When the Raman spectra of gases or vapors are examined, rotational envelopes are found on either side of the Rayleigh line. These are far more intense than the rotational-vibrational bands. If the envelopes can be resolved, the moments of inertia of the molecules can be estimated.

So far, no laser Raman spectra of the Raman band intensities, corrected for instrument effects, have appeared for any substance in both the gas and liquid states. This is a fertile field of research for the study of the liquid state.

2.12 SAMPLE HANDLING AND ACCESSORIES

Each commercial Raman spectrometer has its own line of sample-handling accessories. These are generally designed to handle liquid, solid and crystal-powder samples. Most liquid samples may be examined as they come from the container. Suspended material if present, must first be filtered or centrifuged out. The liquid sample cell is simply a container with rectangular windows, which may be illuminated from one direction and viewed from another. If a liquid is highly colored, or if only microliter or nanoliter amounts are present, it is best examined by the transillumination technique. The liquid is introduced into a capillary tube blown from special non-fluorescent glass. It may either be illuminated through a small, hemispherical lens, blown into the bottom of the capillary (Fig. 2.13a), or transversely (Fig. 2.13b). This technique is also useful for small single crystals or small amounts of crystal powders.

An alternative method of viewing crystal powder samples is shown in Figure 2.13c. The sample is tamped into a well and obliquely illuminated with a focused laser beam. Blocks of polymer may be examined in this way, by cutting a diagonal face into the sample.

Small single crystals may be cemented to the end of a wire and positioned in the laser beam with a micrometer screw. The SPEX Ramalog has a eucentric Goniometer Head accessory, for this purpose.

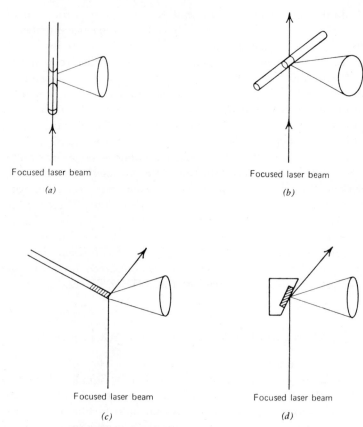

Focused laser beam   Focused laser beam

(a)   (b)

Focused laser beam   Focused laser beam

(c)   (d)

Fig. 2.13 Methods of illuminating microsamples.

When working with small samples, there is sometimes difficulty in getting the sample placed at the intersection of the laser beam focus and slit image. It is desirable to have a laser beam positioner to do this. A laser beam positioner permits the position of the laser beam focus to be moved independently along the three coordinates axes. One possible method of doing this is described by Tobin,[14] and shown in Figure 2.14.

Fluorescent samples present a special category of problems. If the sample is a liquid, it may be distilled, in the hope that the fluorescence is caused by an involatile impurity. Soluble or fusible solid samples may be zone refined, recrystallized, or passed through a chromatographic column. It is very common for the fluorescent sample to be a solid polymer, glass, or biological material which is not amenable to chemical treatment. The first thing to be tried is to allow the sample to sit in the sample space for an hour, with the laser beam on. Gentle heating in an oven will also sometimes work. This

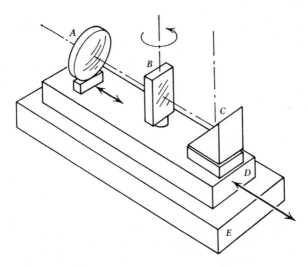

Fig. 2.14 Beam positioner: (*A*) Focusing lens on slide. Motion of slide controls vertical position of focus. (*B*) Parallel-sided plate turning around vertical axis. Rotation of plate controls lateral motion of focus. (*C*) 90° mirror. (*D*) Slide. Motion of slide controls back-and-forth motion of focus. (*E*) Fixed base.

will sometimes volatilize the fluorescent impurity. If this fails, the next thing to do is to locate, if possible, a Raman band above the fluorescence background. The constant a of equation 2.3.6 of this chapter should be determined for a number of exciting wavelengths. If this constant depends on wavelength, one next determines, using equation 2.3.6 of this chapter, which exciting line gives the best signal-to-noise ratio for a given scanning time per spectrometer bandpass. The widest spectrometer slits possible should be used. Once these variables have been optimized, all that can be done is to scan the spectrum slowly enough to get acceptable signal-to-noise ratios. It is often found that a is not very sensitive to wavelength in the 4880–6328 Å range. In this case, the best results will be had with Argon 4880 or 5145 Å, since these lines give the greatest N.

Examination of samples at high or low temperature, or at high pressures, presents special problems. The first difficulty is that when input optics are optimally matched to the spectrometer, the sample is only a few centimeters from the pickup lens. At the cost of signal-to-noise ratio, this may be overcome by using a longer focal length, less than optimal pickup lens. Provided that the sample temperatures are not too high, the pickup lens may be protected by blowing cool, dry nitrogen over it.

One simple means of heating a Raman sample to several hundred degrees centigrade is shown in Figure 2.15*a*. A simple low-temperature accessory,

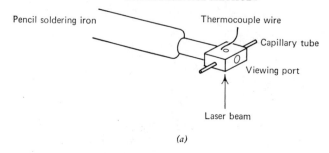

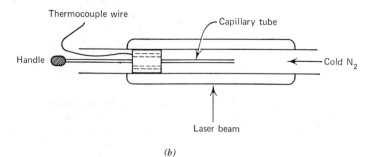

Fig. 2.15   High- and low-temperature sample accessories for Raman spectroscopy.

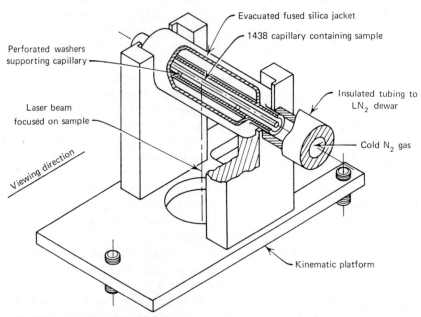

Fig. 2.16   1449 Harney-Miller Variable Temperature Assembly (SPEX Industries, Metuchen, N.J.).

which can be easily built by a glass-blower, is shown in Figure 2.15*b*. A commercial variable temperature cell is available from SPEX Industries, Metuchen, N.J. (Fig. 2.16).

Work at very high temperature or at liquid helium temperatures presents special problems. Coderg (Clichy, France) makes a liquid helium cryostat which may be adapted to Raman spectroscopy. High temperature assemblies must be designed on an *ad hoc* basis.

High-pressure studies may be carried out using an accessory offered by Coderg, or using the diamond cell of Lippincott.[31]

Some limited work in Raman spectroscopy has been done on the spectra of molecules absorbed into silica gel.[32]

2.13   TROUBLE SHOOTING

Under ordinary circumstances, a well-designed instrument will give troublefree operation. Nevertheless, resistors do sometimes burn out, photomultiplier tubes go bad, or optics are knocked out of line. Repair of an amplifier and realignment of a monochromator is a job for a specialist. It is helpful, however, particularly if a component must be shipped out for repair, to know which component to ship.

A commercial instrument should be so designed that the two monochromators can be opened and viewed, the entrance and exit slits can easily be viewed and the various electronic modules are accessible. It is very rare for the spectrometer drive mechanism or electronics to give any trouble. Malfunction is almost invariably either in the optical alignment or in the detector and associated electronics.

A very common cause of trouble is stray noise. This is caused by a high-frequency source in the vicinity of the spectrometer. If the source (which may be the laser) can be located, it can be shielded in a grounded wire cage. If not, an attempt must be made to shield the spectrometer, or at least the electronics. Sometimes a grounded mu-metal shield around, *but not touching* the envelope of the photomultiplier tube will help. An isolation transformer between the power plug and the line may prove useful.

Suppose that the trouble is that no spectrum is recorded, when everything is turned on and all the fuses have been examined. In this event, the photomultiplier voltage is turned off and a turbid liquid put in the sample cell. *Never expose a photomultiplier tube to room lights or to the Rayleigh line when the voltage is on.* Expose it to light as little as possible under any conditions. The beam positioner is adjusted until it is certain that the image of the focal region of the laser beam is falling on the entrance slit. The first monochromator is then opened to assure that both mirrors and the grating are filled with light. The spectrometer is turned to the frequency of the exciting line and the second monochromator opened and examined. If the mirrors and gratings are filled with light, a sheet of black paper is placed so as to protect the photomultiplier tube and the exit slit opened. If light is coming through the exit slit, the trouble does not lie in the monochromator. Some monochromators have a provision for adjusting the second grating. If the trouble is that light misses the exit slit, the grating should be adjusted to correct this.

Supposing the monochromator to be alright, the photomultiplier must next be tested. This is done by feeding the phototube output directly through a 50 Ω terminator into a sensitive oscilloscope, such as the Tektronix 585. The dark pulses should appear on the screen when the phototube voltage is turned on. If no pulses appear and a voltmeter shows that the power supply is providing high voltage to the phototube, then the trouble lies in the phototube or in the dynode chain. If pulses do appear, the photomultiplier voltage should be turned off, a 2.0 or 3.0 neutral density filter put after the exit slit, and the output optics checked to make sure that light can reach the photocathode. With a 3.0 neutral density filter in place, turn on the photomultiplier voltage and quickly scan the monochromator past the Rayleigh line. If a burst of pulses appears on the oscilloscope screen as the Rayleigh line goes past the exit slit, the trouble must lie in the external electronics. With the photomultiplier connected to the pulse amplifiers, the output of the amplifiers and the pulse shaper module should be examined in turn with the oscilloscope. If these are all operating properly, the output of the RC integrating circuit should be tested with an electrometer, while a signal is being fed into the photomultiplier tube. The d-c electrometer provided with the instrument for d-c recording may be suitable for this purpose. Lastly, the recorder may be put on a suitable range and tested with a dry cell put across its input.

The foregoing systematic procedure should enable one to locate quickly a defective component. An apparent malfunction of Raman spectrometers will appear if the photomultiplier tube is accidently exposed to the Rayleigh line with the voltage on. If this happens and the exposure was short, the voltage should be turned off and the tube allowed to rest in complete darkness for 12 hours.

Schematic drawings of electronic circuits are sometimes provided with an instrument. If a good electronics shop is at hand, they may be able to repair a given electronic module. The individual worker should not attempt this unless he is highly skilled in electronics. One should, of course, carefully read any instruction manual or literature provided by the manufacturer.

THE INTERPRETATION OF RAMAN SPECTRA

3.1 INTRODUCTION TO THE CONCEPT OF GROUP FREQUENCIES

RAMAN SPECTROSCOPY, INFRARED ABSORPTION SPECTROSCOPY, AND MOLECULAR STRUCTURE

The preceding two chapters have outlined the theory and practice of Raman spectroscopy, in general terms. We must now turn to the problem of analyzing a spectrum, once it has been recorded. At this point, a discussion of the relation between Raman spectroscopy and infrared absorption spectroscopy is in order. Like the Raman effect, infrared absorption is due primarily to vibrational and rotational transitions. The determining quantity is the transition moment.

$$M = \int u_j^* \mu u_i \, dv \qquad (3.1.1)$$

The dipole moment operator, μ, has the symmetry properties of the normal coordinates, rather than of their products. We thus expect and find, differences in the selection rules between the two kinds of spectrum. The normal frequencies deduced for a given material will, of course, be the same from either spectrum. One might reasonably ask, under what circumstances would it be preferable to record the infrared spectrum rather than the Raman spectrum, or vice versa? The answer would seem to be that, for routine work in organic chemistry laboratories, infrared absorption spectroscopy is the method of choice. In main, this is a matter of economics, since a number of companies sell small, tabletop infrared spectrometers for a few thousand dollars.

As soon as a problem becomes nonroutine, that is, if one is interested in molecular structure determinations, the study of proteins, unusual quantitative analysis, and so on, then it is essential to have high-quality infrared *and* Raman spectra available. If one wishes to study the effect of some kind of treatment on the structure of polymer films, one would like to be able to record the infrared spectrum, using polarized radiation and the Raman spectrum. As examples of cases where Raman spectroscopy might be the method of choice, the study of aqueous solutions and the recording of spectra of the output of a gas chromatograph, might be cited. One might add that

79

a Raman spectrum covers the entire region from 10 cm^{-1} to 4000 cm^{-1} in a single run. Recording the infrared spectrum, in the region from 10 cm^{-1} to 200 cm^{-1}, requires expensive equipment and is time-consuming.

This having been said, we next turn to some general remarks on the determination of molecular structure from vibrational spectra. This subject has been treated in a number of excellent standard texts and has been exemplified in literally thousands of articles. The reader is referred to the bibliography for the names of texts where details may be found. Through a careful study of the infrared and Raman spectra of a material, including the rotational and rotational-vibrational spectra of the vapor, it is sometimes possible to deduce the geometrical shape of the molecules making up the material. Candidates for this kind of analysis are molecules with fewer than twenty atoms and which are expected to have some elements of symmetry. The overall procedure is to write down possible structures and to predict from these the expected infrared and Raman activities of the normal modes, the expected rotational-vibrational band contours and some characteristic chemical group frequencies. These predictions are compared with the observed spectra, and the structure with the best fit selected. Moments of inertia can be estimated from the rotational-vibrational spectra of gases, to get an estimate of the size of the molecule. Often, just the analysis of the selection rules will give the answer, although a few cases are known where this limited approach has given a wrong structure.

For large, unsymmetrical molecules, one is forced to rely almost completely on a table of chemical group frequencies[R10, R11, R12] to get some partial information about molecular structure. This is to say, one tries to identify observed infrared or Raman bands with specific chemical groupings in a molecule. While a complete determination of the structure is out of the question in this case, it it may be that the structure is known, from other considerations, to be one of two or three possible ones. An examination of group frequencies will often make it possible to select one of these. If the sample is a polymer, which can be orientated, or a sample which can be obtained as a single crystal, additional information can be gotten from the infrared spectrum, using polarized radiation and from the Raman spectrum.

One might reasonably, ask why not just do an X-ray diffraction analysis and get a complete, unambiguous structural determination? Indeed, for problems of major importance, this is what is done. An X-ray diffraction analysis, however, suffers from two major drawbacks compared to a spectroscopic study. The first is that both the experimental work and the subsequent computations are very lengthy. More serious, X-ray structure determinations can only be carried out on well-formed single crystals. The method is helpless when the sample is a liquid or a gas. Vibrational spectroscopy is equally useful for any of the states of matter.

The group frequency charts which have been developed for infrared spectroscopy generally contain some data for the analysis of Raman spectra. A useful fact is that for some group frequencies, the associated infrared bands are weak and the Raman bands strong, or vice versa. It is to be regretted that quantitative intensity data of this sort are not widely available for Raman spectra, as they are for infrared spectra. Hopefully, this lack will be repaired in the future. Until very recently, group frequency charts covered only the region 700–4000 cm^{-1}. A new book[R12] charts the region 300–700 cm^{-1}. The region from 0–300 cm^{-1}, which is readily examined in the Raman spectrum, still remains to be analyzed.

It is worth remarking that as a practical matter, Raman spectroscopy is one of the best ways of making a qualitative identification of refractory samples, like glasses, polymers, grits and the like. This is because a spectrum may be recorded with oblique front illumination, without any necessity for sample preparation.

At the time of the writing of this book, laser Raman spectrometers are just coming into the hands of chemists. The current output of published work is devoted primarily to the elucidation of molecular structure or the study of crystal physics. The major exceptions are some work on the identification of natural products and of polymers. The whole field of quantitative analysis, particularly in aqueous solution, is virtually untouched. For this reason, this chapter and the next will deal primarily with molecular structure determination. However, the potential application of results to analytical chemistry will be underlined, in each case.

BASIS OF THE CONCEPT OF GROUP FREQUENCIES

As was mentioned in Chapter 1, equation 1.3.3 of that chapter can be recast into such a form that the force constants along the diagonal of the secular determinant refer to the stretching, bending or torsion of chemical groupings. The coordinates to which these force constants refer are called "internal coordinates."[R3] One may reasonably refer to an SH stretching force constant, to an HCH bending force constant, to an NO_3^- out-of-plane bending force constant, etc. It is an experimental fact that these force constants are often much the same for different molecules. A carbon-carbon double bond will have a stretching force constant of about 9.7×10^6 dyne cm^{-1} regardless of the molecule in which it occurs. Extensive tables of force constants are now available[R3] so that rough calculations of the vibrational spectra of a molecule whose structure is known can be made *ab initio*. The reason for this is that the off-diagonal force constants which appear in the secular equation are generally small. It is this empirical fact which makes it reasonable to talk about group frequencies. There are, of course, individual molecules for which some of the off-diagonal force constants are not small. In these cases, some group frequencies may be atypical. It often happens, particularly for molecules which contain heavy atoms, that the group frequencies for different chemical groupings will not be clearly distinguishable from one another. This happens, for example, in metalorganic compounds, for which CH_3—metal stretching frequencies fall into a relatively narrow range. One may cite the tetramethyl compounds of lead, tin and germanium tetramethyls[33] in which the skeletal metal-carbon stretching frequencies (two to each compound) lie at (459, 478), (507, 532) and (558, 595) cm^{-1}, respectively.

There are certain empirical approximations which prove to be helpful in making use of group frequencies. In trimethyl tin iodide,[34] as in similar compounds containing heavy atoms with light side groups, the spectrum may be considered to be a superposition of frequencies characteristic of a CH_3 group, frequencies characteristic of the C_3SnI skeleton and mixed frequencies. The latter are, in effect, restricted rotations of a methyl group against the rest of the molecule. The assignment of frequencies for $(CH_3)_3SnI$

Table 3.1. The Assignment of Frequencies for $(CH_3)_3SnI$, cm^{-1}

	Frequency cm^{-1}	Assignment	Type of Vibration
1	2992	A_1	$\nu(CH_3)$
2	2917	A_1	$\nu(CH_3)$
3	1328	A_1	$\delta(CH_3)$
4	1195	A_1	$\delta(CH_3)$
5	704	A_1	CH_3 rocking
6	511	A_1	$\nu(Sn-C)$
7	177	A_1	$\nu(Sn-I)$
8	149	A_1	$\delta(SnC_3)$
9	(2992)	A_2	$\nu(CH_3)$
10	(1402)	A_2	$\delta(CH_3)$
11	(795)	A_2	CH_3 rocking
12	—	A_2	CH_3 torsion
13	2992	E	$\nu(CH_3)$
14	2992	E	$\nu(CH_3)$
15	2917	E	$\nu(CH_3)$
16	1402	E	$\delta(CH_3)$
17	1328	E	$\delta(CH_3)$
18	1195	E	$\delta(CH_3)$
19	795	E	CH_3 rocking
20	704	E	CH_3 rocking
21	538	E	$\nu(Sn-C)$
22	149	E	$\delta(SnC_3)$
23	117	E	Rocking
24	—	E	CH_3 torsion

is given in Table 3.1. The six skeletal frequencies (Nos. 6, 7, 8, 21, 22 and 23) a lllie below 600 cm^{-1} and fall into a pattern reminiscent of the spectrum of CH_3Cl. The methyl frequencies are well known from the spectra of methyl compounds and are easily identified. The mixed frequencies must be identified from the intensity and wavenumber values, from the polarization of the Raman bands and a normal coordinate analysis, although an experienced worker can make very good guesses. Another example of the persistence of characteristic frequencies of a side group, the infrared spectrum of polyethylene terephthalate may be cited.[35] (Unfortunately, no Raman spectrum has been reported.) The frequencies characteristic of a para-substituted benzene ring are quite readily picked out of the spectrum. The assignment is shown in Table 3.2. This table is instructive for another reason. It is a

Table 3.2. Assignments of Frequencies for Dimethyl Terephthalate, Polyethylene Terephthalate, and Poly-p-xylene, cm^{-1}

Benzene	p-Dichloro-benzene	Dimethyl Terephthalate	Polyethylene Terephthalate	Poly-p-xylene	Mode	Symmetry Species
		Raman Ring Frequencies				
3062	3079	3050			2	A_g
3047	3079	3050			7b	B_{3g}
1597	1630	1609			8b	B_{3g}
	1576	1609			8a	A_g
1340	1331	1262			3	B_{3g}
1178	1109	1104 or 1169			9a	A_g
992	1070	1104			1	A_g
995	942	1004			5	B_{2g}
	748	930^a			7a	A_g
849	855	814			10a	B_{1g}
703	710	705			4	B_{2g}
606	627	630			6b	B_{3g}
	355	266			9b	B_{3g}
	333	266			6a	A_g
	302	172			10b	B_{2g}
		Infrared and Inactive Ring Frequencies				
3060	3060	3060	3040	3060	13	B_{1u}
3080	3060	3060	3040	3060	20b	B_{2u}
1648 or 1310	1624 or 1345	1676	1685	1680	14	B_{2u}
1485	1475	1505	1508	1510	19a	B_{1u}
	1400	1410	1410	1415	19b	B_{2u}
1010	1115	1105	1140	1138	12	B_{1u}

Table 3.2 (*Continued*)

Benzene	p-Dichloro-benzene	Dimethyl Terephthalate	Polyethylene Terephthalate	Poly-p-xylene	Mode	Symmetry Species
			Infrared and Inactive Ring Frequencies			
1110 or				1080–		
1152	1096	1090	1108	1100	15	B_{2u}
1037	1018	1024	1018	1018	18a	B_{1u}
	780	874	873	860	20a	B_{1u}
	824	735	727	820	17b	B_{3u}
673	550	498	500	543	11	B_{3u}
	488	322	382	?	16b	B_{3u}
	200	200			18b	B_{2u}
975	950	965			17a	A_u
405	454	400			16a	A_u
		Raman External Frequencies				Assignment[b]
		1729				$\nu(C{=}O)$
		1291(1296)				$\nu(COC)$
		1104(?)				$\nu(COC)$
		336 or 366(?)				$\nu(COC)$
		2982				$\nu(CH)$
		2940				$\nu(CH)$
		1447				$\delta(CH_2)$
		1397				$\delta(CH_2)$
		1361				$\delta(CH_2)$
		336 or 366(?)				$\delta(CCO)$
		Infrared External Frequencies				
		1735	1725			$\nu(C{=}O)$
		1275	1230–80			$\nu(COC)$
		1120	1140			$\nu(COC)$
		435	422			$\delta(COC)$
			2960			$\nu(CH)$
			2900			$\nu(CH)$
			1448			$\delta(CH_2)$
			1368			$\delta(CH_2)$
			1340			$\delta(CH_2)$

[a] Fermi resonance doublet with $200 + 705$.
[b] ν = stretching, δ = deformation.

consequence of the preceding remarks, that if the vibrational spectra of a series of structurally similar compounds is examined, patterns will emerge.[R13] It is feasible to trace a spectral band, arising from a given normal mode of vibration, through the series. The Raman spectrum is particularly helpful for this kind of work, since the totally symmetric normal modes can often be picked out by their high degree of polarization.

It is to be regretted that more work has not been done with amorphous polymers to establish group frequencies. If such a polymer is stetched, the various infrared dipole moments and Raman polarizations will have a preferred orientation in space. By examining the direction of the dipole moment and polarization vectors of normal modes, close-lying bands can often be assigned to one normal mode or the other. Another useful approach is to bring about a chemical change in the molecule.

We now turn to a discussion of the cases where the group frequency approach may prove deceptive. A bare table of characteristic group frequencies is of limited value. Some additional information is always given. Carbonyl stretching bands, for instance, tend to be strong in both the infrared and Raman spectra and are of moderate breadth. Double-bond $C=C$ stretching frequencies tend to be of weak-to-moderate strength in the infrared and to have very narrow bandwidths. They fall near 1600 cm^{-1}. They tend to be very strong and polarized in the Raman spectrum. There is generally a weak, sharp CH stretching frequency near 3100 cm^{-1}. The presence of these two bands is generally a good indication of carbon-carbon double bonds, although aromatic rings give a similar pattern. (A distinction is possible by examining the remainder of the infrared and Raman spectra. See below.) Now, if only one double bond is present in a long hydrocarbon chain, the 3100 cm^{-1} band might not be visible and the $C=C$ stretching frequency would be weak. Bands characteristic of the double bond, in the remainder of the spectrum, would probably be weak and indistinguishable from the carbon-chain bands. In this kind of case, there would be some danger of confusing the double-bond frequency with a carbonyl frequency, (Were the Raman band of much greater strength relative to the rest of the spectrum, than in the infrared spectrum, one would suspect a double bond rather than a carbonyl frequency.)

Another troublesome case arises in the vibrational spectra of molecules made up of a large number of chemical components, like proteins or deoxyribonucleic acid. These spectra are, to some extent, a superposition of the spectra of a number of simpler components. A spectroscopist given the spectrum of a protein could tell that he was dealing with a hydrogen-bonded amide, but not much else.

It happens very commonly that two characteristic group frequencies will fall close to one another. For instance, a $C-S$ stretching frequency may appear, in the Raman spectrum, near 600–630 cm^{-1} and a C—Cl stretching frequency near 620 cm^{-1}. With no other information available, it might be hard to distinguish these from one another. If one wants the structural information badly enough, then the vibrational spectra may be supplemented with other data. For instance, an elemental analysis will eliminate many possibilities. Or, the deuterium analog of a new compound may be synthesized. This will shift CH bands, but leave other bands relatively unchanged.

The foregoing is not intended to frighten a would-be user away from the group frequency concept. An experienced vibrational spectroscopist can deduce a great deal of valuable information from the spectra of an unknown compound. It is worthwhile, however, to sound a cautionary note which is not often heard in the standard texts.

We next turn to a consideration of the factors which influence the intensities of infrared and Raman bands.* It is a good rule of thumb that polar molecules tend to be strong infrared absorbers and weak Raman scatterers. Nonpolar molecules tend to be weak infrared absorbers and may be strong Raman scatterers. These rules are valid if applied to a vibrational mode associated with, say, a polar bond in an otherwise nonpolar molecule. The degree of polarity of a bond is given by[36]

$$z = 1 - \exp - \frac{(x_A - x_B)^2}{4} \tag{3.1.2}$$

where x_A and x_B are the electronegativities of the atoms making up the bond. A table of electronegativities is given by Hannay and Smythe[37] as well as by Gordy.[38]

Electronegativity is not a use predictor of Raman intensities if bonds of different order are being compared. For a given bond, the Raman intensity of the stretching band associated with this bond will rise rapidly with bond order. Gordy's formula for bond order is

$$N = \frac{a}{R^2} + b \tag{3.1.3}$$

where R = bond length, Å. Electronegativities are given in Table 3.3a, the constants a and b in Table 3.3b.

There is some limited evidence[R4] that for a given type of vibration in a series of compounds with analogous structure, for example XY_4 compounds, the Raman intensity goes as

---

*The remainder of this chapter is adapted from References R10, R11, and R14, with additions by the author. The author wishes to thank Dr. H. A. Szymanski for permission to use tables and a figure from his text, as well as for providing a copy of his text.

Table 3.3a. Electronegativities of the Elements[38]

Ag 1.9		
Al 1.5	Ga 1.4	S 2.53
As 2.0	Ge 1.7	Sb 1.8
Au 3.1	H. 2.13	Sc 1.3
B 1.9	Hg 1.0	Se 2.4
Ba 0.9	I 2.45	Si 1.8
Be 1.45	In 1.4	Sn 1.7
Bi 1.8	K 0.80	Sr 1.0
Br 2.75	Li 0.95	Te 2.1
C 2.55	Mg 1.2	Ti 1.6
Ca 1.0	N 2.98	Y 1.3
Cd 1.1	Na 0.90	Zn 1.2
Cl 2.97	O 3.45	Zr 1.6
Cs 0.75	P 2.1	
Cu 2.2	Pb 1.5	
F 3.95	Rb 0.78	

Table 3.3b. Constants for Bond Order Determination

Bond	a	b
BB	9.12	-1.94
BC	8.05	-2.11
BN	7.15	-2.10
BO	6.75	-2.14
CC	6.80	-1.71
CN	6.48	-2.00
CO	5.75	-1.85
CS	11.9	-2.59
NN	5.28	-1.41
NO	4.98	-1.45
OO	4.73	-1.22

$$I \propto (1 - z) \tag{3.1.4}$$

where z is given by equation 3.1.2.

The dependence of intensity on bond order is much greater. A considerable amount of effort has gone into trying to correlate Raman intensities or the change of polarizability with the change in a normal coordinate, with bond order, etc. The results have been, by and large, disappointing. This is not surprising, because even if such a relation exists, the intensities and normal coordinates are poorly known. Some detailed rules of thumb are the following:

1. Unsaturated bonds like $-C=C-$, $-C\equiv N$, $-C\equiv C-$, etc., tend to give rise to very strong Raman bands and to weak infrared bands between 1600 cm^{-1} and 2300 cm^{-1}.

2. Bonds containing one or more heavy atoms, like $-CI$, $Pb-Pb$, etc., will give rise to extremely strong Raman scattering and, sometimes, to strong infrared absorption.

3. The stretching frequencies of the polar $-NH$ and $-OH$ bonds give rise to very strong infrared absorption, but weak Raman scattering.

4. The $-OH$ and $-NH$ bending bands are generally weak in the Raman spectrum, but strong in the infrared spectrum.

5. The CH bending frequencies are generally considerably weaker than the CH stretching frequencies, in the Raman spectrum.

6. The methyl and methylene bending frequencies are strong in the infrared spectrum, but tend to be weak in the Raman spectrum.

7. The $-SH$ and $-S-S-$ stretching frequencies are strong in the Raman spectrum, weak in the infrared spectrum.

8. The $-C-C-$ stretching bands are strong in the Raman spectrum, weak in the infrared spectrum.

9. The $C-H$ stretching frequencies in $-C=C{<}^{H}_{H'}$, $-C\equiv CH$, etc., are strong in the Raman spectrum, weak to moderate in the infrared spectrum

10. Carbonyl and carboxyl frequencies tend to be strong in both the infrared and Raman spectra.

11. Aromatic compounds tend to have strong infrared absorption and strong Raman scattering.

12. All ring compounds, saturated or aromatic, have a very strong Raman band associated with the symmetric expansion and contraction of the rings.

13. With this exception, ring deformation vibrations are weak in the Raman spectrum and strong in the infrared spectrum.

14. Out-of-plane bending bands of olefines and aromatics tend to be strong in both the infrared and Raman spectra. However, a given band is seldom strong in both.

15. For groups like $O=C=O$, $C-O-C$, $N=C-N$, etc., there will be one symmetric and one asymmetric stretching frequency. The former will generally be lower in frequency than the latter and will be strong in the Raman spectrum and weak in the infrared spectrum. The asymmetric frequency will tend to be weak in the Raman spectrum and strong in the infrared spectrum.

16. Polynuclear aromatic compounds will tend to have complex infrared and Raman spectra composed of strong, sharp bands.

17. In hydrogen-bonded compounds, the —OH and —NH stretching and bending frequencies are very broad. Hydrogen bonding may produce shifts in amide or carbonyl stretching frequencies.

One should keep in mind that there are occasional exceptions to all of these rules.

It was noted in Chapter 1 that for compounds having a center of symmetry, no band can appear in both the infrared and Raman spectrum. There may, however, be near coincidences. In *trans-dicyanothylene*, one of the two C≡N stretching bands appears in the infrared spectrum and one in the Raman spectrum. Their frequencies are quite close. A good rule of thumb is that in centrosymmetric molecules, the more spatially separated like chemical groups are, the closer together the corresponding frequencies will be. Even a bond in a symmetrical environment in a noncentrosymmetric molecule will show effects of a pseudomutual exclusion rule. Of the two C—Cl stretching frequencies of *trans*-4,5-dichloro-4-decene, one will be strong in the Raman and the other in the infrared.

As was noted earlier, the group frequency correlation tables available for Raman spectra do not approach in completeness those available for infrared spectra. One of the best compilations, that of Brandmüller and Moser[R2] is reproduced in Tables 3.4a and 3.4b. Also given at the end of the table is

Table 3.4a. Group Frequencies in the Raman Spectrum^a

4400	HH valence	Hydrogen	
3636–3150	OH valence	Hydrates	Broad, one or two bands
3374	≡CH symmetrical valence	Acetylene	
3372	NH valence	$RH_2 \cdot C\,NH_2$	Also at 3310
3356	NH valence	$RR'HC \cdot NH_2$	Also at 3307
3350	C=NOH	RHC=NOH and RR'C=NOH	Also at 1600
3335	NH valence	C=NH	
3330	NH valence	R_2NH	
3310	NH valence	$RH_2C \cdot NH_2$	Also at 3372
3307	NH valence	$RR'HC \cdot NH_2$	Also at 3356
3305–3270	≡CH valence	Monosubstituted acelylenes	IR
3108	=CH valence	Ethylene	
3070–3045	=CH valence	Mono-, di-, and trisubstituted aromatics	
3062, 3047	=CH valence	Benzene	

Table 3.4*a* (*Continued*)

3019	=CH valence	Ethylene	
3000–2800	CH	Aliphatic molecules	
2790	CH	>N·CH$_3$	
2570	SH	RSH	
2329	C≡N valence	Nitriles	
2304	C≡C valence	Dialkylacetylenes	Second band at 2227
2300	Se—H	R·Se·H	
2252	RC≡CC≡CR	Completely substi- tuted diacetylenes	
2250	C≡N valence	Unconjugated nitriles	IR
2227	C≡C valence	Dialkylacetylenes	Second band at 2304
2225	C≡N valence	Conjugated nitriles	IR
2187	Si—H symmetrical valence	Silane	
2180	C=N	RN=C=S	Also at 2105
2150	C≡N	RSC≡N, RN≡C	
2125–2118	C≡C valence	Monoalkyl acetylenes	
2105	C=N	RN=C=S	Also at 2180
2104	N=N≡N	H$_3$C·N=N≡N	
2089	C≡N valence	Nitriles	IR
2049	C=C=O valence	Ketenes	
1980	C=C=C assymmetrical valence	Allene and derivatives	Strong in IR
1974	C≡C valence	Acetylene	
1820–1650	C=O valence	Carbonyl compounds	IR
1804, 1745	C=O	R·CO·O·CO·R′	
1792	C=O	R·CO·Cl	
1776	C=O	Cl·CO·OR	
1734	C=O	R·CO·OR′	
1720	C=O	R·CO·H	
1710	C=O	R·CO·CH$_3$	
1690	C=O	NH$_2$·CO·OR	
1675	C=O	R·CO·NH$_2$	
1652	C=O	R·CO·OH	
1680–1640	C=C valence	Olefines and derivatives	Weak in IR
1676	C=C	R$_2$C=CR$_2'$	
1674	C=C	RHC=CHR′ trans	
1670	C=N	RHC=NR′	
1665	C=N	RR′C=NOH	
1660	C=NOH	R·HC=NOH, RR′C=NOH	Also at 3350
1658	C=C	RHC=CHR′cis	
1654	C=N	R·(OR)C=NH Ar·HC=NR	
1642	C=C	RHC=CH$_2$	

Table 3.4a (*Continued*)

1640	O—N=O valence	Nitric acid esters	Methyl esters of nitric acid also at 1603
1630	C=N	Ar·HC=NH	
1623		Nitric acid esters	Also at 570, 1285–1260
1621	C=C	Ethylene	
1610	N=O	R·NO	
1610–1590	Ring vibration	Aromatic hydrocarbons	IR
1603	O—N=O valence	Methyl ester of nitric aicd	Also at 1640
1576	N=N valence		Weak in IR
1571	C=C valence	$Cl_2C=CCl_2$	Very weak in IR
1570–1520	N—NO_2 asymmetrical valence	Alkyl nitramines	IR
1550	NO_2	RNO_2	Also at 610, 1380
1520	NO_2	$ArNO_2$	Also at 535, 1340
1500		Cyclopentadiene, furane, pyrrole, and their derivatives	
1460	CH_3 asymmetrical deformation	Aliphatic hydrocarbons	IR
1455	CH_2 bending vibration	Cyclopentane derivatives	IR
1452	CH_2 bending vibration	Cyclohexane derivatives	IR
1450–1200	C—C—H in-plane CH deformation	Substituted ethylenes	Weak in IR
1434–1409	N—C—O symmetrical valence	Alkyl isocyanates	
1411	—SO_2 asymmetrical valence	Cl_2SO_2	
1393–1373	CH_3	CH_3 on ring	
1390	NO_3^- vibration	Nitrate ion	
1380	—SO_2 asymmetrical valence	$RO·SO_2OR$	
1380	NO_2	$R·NO_2$	Also at 610, 1550
1380		Naphthalene, cyclopentadiene, furane, pyrrole, and derivatives	
1370	N—NO_2 symmetrical valence	Alkyl nitramines	
1360–1330	CH deformation	Branched hydrocarbon chains	IR

Table 3.4a *(Continued)*

1350–1150	CH$_2$ wagging and torsion		IR torsion (weak)
1340	NO$_2$	Ar·NO$_2$	Also at 535, 1520
1340–1310	N—NO$_2$ symmetrical valence	Alkyl nitramines	IR
1331	NO$_2^-$	Nitrite ion	
1330–1313	Ring vibration	1,3-Diaklyl-substituted aromatics	
1305	CH$_2$ wagging	Linear hydrocarbon chains	Weak in IR
1300		1,3-Disubstituted aromatics with longer side chains	
1285–1260	O—NO$_2$ symmetrical valence	Nitric acid esters	IR Also at 570, 1623
1276	N=N≡N	H$_3$C·N=N≡N	
1275–1200	COC Asymmetrical valence	C=C—O—C In ethers, etc.	IR
1270	SO$_2$ asymmetrical valence	R′SO$_2$R	
1268	Symmetrical ring vibration	Ethylene oxide	IR
1250	Methyl rocking(?)	Tertiary butyl groups	IR. Also band at 1210–1200
1240	NO$_2^-$	Nitrite ion	
1230	S=O valence	Cl·SO·Cl	
1216	S=O valence	Cl·SO·OR	
1210–1200	Methyl rocking	Tertiary butyl groups	IR. Also band at 1250
1208–1183		Monoalkyl and 1,4-dialkyl substituted aromatics	
1205–1125	COH valence	Saturated aliphatic tertiary alcohols, highly symmetrical secondary alcohols	IR
1200	S=O valence	RO·SO·OR Alkyl ketones	
1200–1187		1.3-Dialkyl substituted aromatics	
1190	SO$_2$ symmetrical valence	RO·SO$_2$Cl, RO·SO$_2$·OR, Cl·SO$_2$Cl	
1186	SO$_2$ symmetrical valence	Cl·SO$_2$Sl	

Table 3.4a (*Continued*)

1186	Ring vibration	Cyclopropane derivatives	
1178–1152		Monosubstituted aromatics	
1175	Asymmetrical ring vibration	1.3.5-Trioxanes	Strong in IR
1170	Methyl rocking(?)	Isopropyl groups	IR
1130	SO_2 symmetrical valence	$R \cdot SO_2R$	
1130–1120	$C{=}C{=}O$ symmetrical valence	Ketenes and their alkyl derivatives	Strong in IR
1125–1085	Asymmetrical ring vibration	1,4—Dioxanes	IR
1125–1085	C—OH valence	α-Unsaturated or cyclic tertiary alcohols, saturated secondary aliphatic alcohols	Strong in IR
1100–800	CC chain	n-Paraffiines	
1100–670	CC chain	Branched paraffines	
1085	C—OH valence	α-Unsaturated secondary alcohols, n-alkyl primary alcohols, acyclic secondary carbinols with 5- or 6- membered rings	Strong in IR
1075–1020	COC symmetrical valence	$C{=}C{-}O{-}C$ in vinyl and aryl ethers, etc.	IR
1070	$C{=}C{=}C$ symmetrical valence	Allenes and their alkyl derivatives	Strong in IR
1050	NO_3^-	Nitrate ion	
1050	C—OH valence	Diunsaturated secondary alcohols, α-branched chains, and/or unsaturated primary alcohols, highly unsaturated tertiary alcohols	Strong in IR
1050–1030	Ring vibration	1.2-Disubstituted aromatics	
1048	C—F	$H_3C{-}F$	
1046	N=H	Aromatic azo compounds	Weak in IR
1035–1017	Ring vibration	Monosubstituted aromatics	
1030	S=O valence	$R \cdot SO \cdot R$	IR

Table 3.4a (*Continued*)

1028	Symmetrical ring vibration	Trimethylene oxides	IR
1007–990	Ring vibration	Benzene, mono-,1,3-disubstituted and 1,3,5-trisubstituted aromatics	
980–971	Asymmetrical ring vibration	Trimethylene oxides	
970	Ring vibration	Cyclobutane	
958	Symmetrical ring vibration	1,3,5-trioxanes	IR
930	C—C valence(?)	Tertiary butyl group	Weak in IR
930–910	ClO_3^-	Chlorate ion	Weak in IR
918	C—O—C symmetrical valence	Dimethyl ether	
913	Symmetrical ring vibration	Tetrahydrofurane	IR
899–884	Ring vibration	Mono-,1,1-di-,1,2-di-, 1,1,2-trisubstituted cyclopentanes	
880	C—N valence	Aliphatic nitro compounds	IR
877	O—O valence	H_2O_2, organic peroxides	Weak in IR, not present for all peroxides
876	N—N	$H_2N\,NH_2$	
850–820	Ring vibration	Monosubstituted aromatics with simple substituents (CH_3, NH_2, Cl, etc.)	—
835–795	C—C valence(?)	Isopropyl groups, tertiary butyl groups	Weak in IR
832	S—H in-plane deformation	Ethyl mercaptan	
822	Ring vibration	Cyclohexane (boat)	
813	NO_2^-	Nitrite ion	
813	Symmetrical ring vibration	Tetrahydropyrane	IR
802	Ring vibration	Cyclohexane (chair)	
770		β-Substituted naphthalenes	
742–716	CH deformation	1,2,4-Trisubstituted aromatics	
736–711	CH deformation	1,2-Disubstituted aromatics	
733	Ring vibration	Cycloheptane	

Table 3.4a (*Continued*)

721–711	CH deformation	1,3-Disubstituted aromatics	
710	CCl	$H_3C \cdot Cl$	
705–685	CS valence	CH_3S	IR
703	Ring vibration	Cyclooctane	
700–600	C—SH valence	Mercaptans	Weak in IR
660–630	CS valence	RCH_2S	
652	CH deformation	1,2,3-trisubstituted aromatics	
650–600	C≡CH deformation	Monosubstituted acetylenes	IR, overtone at 1260
650	CCl	$RH_2C \cdot Cl$	
640	Ring vibration	1,3-Disubstituted aromatics	
630	HC≡CR	Monosubstituted acetylenes	Also at 340
630–600	CS valence	RR'HCS	IR
623–610	Ring vibration	Monosubstituted aromatics	
620	CH deformation	Monosubstituted aromatics	
610	C—NO_2 deformation	RNO_2	IR at 1380, 1550
610	CCl	R_2HC Cl	
600–570	C—S valence	RR'R''C S	
594	C—Br	$H_3C \cdot Br$	
580		RHC=CHR' cis	Also at 413, 297
570	NO_2	Nitric acid esters	Also at 1285–1260, 1623
570	C—Cl	$R_3C\,C'l$	
570–554	CH deformation	1,3,5-Trisubstituted aromatics	
560	C—Br	$RH_2C'Br$	
550–450	S—S valence	Alkyl disulfides	
535	NO_2	$Ar \cdot NO_2$	Also at 1340, 1520
530	C—Br	$R_2HC \cdot Br$	
522	C—I	$H_3C \cdot I$	
510	C—Br	$R_3C \cdot Br$	
500	C—I	$RH_2C \cdot I$	
490		RHC=CHR' (trans)	Also at 210
490	C—I	$R_2HC \cdot I$	
481	Ring vibration	1,2,3-Trisubtituted aromatics	
480	C—I	$R_3C \cdot I$	
476–465	Ring vibration	1,2,3-Trisubstituted aromatics	
445	S—S	ClS—SCl	
435		RHC=CH_2	
435	Si—Si	H_3Si—SiH_3	
434		RR'C=CH_2	Also at 394, 261

Table 3.4a (*Continued*)

394		RR′C=CH₂	Also at 434, 261
365		Disubstituted acetylenes	
340		Monosubstituted acetylenes	Also at 630
334–234	CC deformation	n-Hydrocarbon chains	Frequency decreases with increasing chain length
332	SH out-of-plane deformation	Mercaptans	
261		RR′C=CH₂	Also at 434, 394
210		RHC=CHR′ (trans)	Also at 490
about 200	R—C—C≡CH, R—C—C≡N	Acetylene and nitrile derivatives with 4-membered and longer chains	

From Brandmüller and Moser,[R2] page 471 *et seq.*
[a] IR = infrared active; R = alkyl; Ar = aromatic.

Table 3.4b. Raman Group Frequencies for Symmetric Ring-Stretching Vibrations.[R14] Position of the Symmetric Ring Stretch in the Raman

Ring System	Band Position
Cyclopropane	1185
Azacyclopropane	1210
Cyclobutane	1000
Ethylene oxide	1268
Cyclopentane	890
Trimethylene oxide	1028
Cyclohexane (boat isomer)[a]	820
Cyclohexane (chair isomer)	800
Tetrahydrofuran	915
Tetrahydropyran	813
1,3,5-Trioxane	960
1,4-Dioxane	835
Cycloheptane	735
Cyclooctane	705
Aromatic Rings	
A. Benzene	995
B. Parasubstituted	860
C. Monosubstituted	1000

[a] For cyclohexanol the CO stretch for the chair and boat form differ by 100 wavenumbers.

the listing of symmetric ring-stretching Raman frequencies given by Szyman-ski.[R14] All data are based on mercury arc excitation, but the band positions are certainly accurate and the qualitative relative intensities are probably correct.

The following section gives a short survey of the application of the group frequency concept to the determination of molecular structure. They are not intended to substitute for the standard works in the field, but only to provide background and a handy reference. Skill in the art of interpreting spectra is only acquired by careful reading of the literature and long experience.

As the reader examines the following sections, he will realize that a structural determination based on vibrational spectra alone is often ambiguous. Amines and alcohols, or olefines and phenyl derivatives, for example, are often hard to distinguish from one another. This need cause little concern. The person submitting a sample for examination will generally have had an elemental analysis done and have a rough idea of what his sample may or may not contain. If the sample is known not to contain nitrogen, for example, then bands which might be alcohol or amine bands must be alcohol bands.

3.2 CHARACTERISTIC GROUP FREQUENCIES

ALKANES

Normal alkanes in the melt or in solution consist of a mixture of a large number of rotational isomers. The spectra of the lower members of the series differ considerably from one another. Chains with more than twenty carbon atoms tend to give infrared and Raman spectra which differ little from compound to compound. This is particularly true if the sample is in the solid state. The prototype long-chain alkane is, of course, polyethylene. The actual spectra of polyethylene are complicated by the fact that it is a mixture of crystalline and amorphous material and that it may have branched chains or pendant side-groups. The task of comparing its spectrum to that of shorter-chain alkanes is complicated by the fact that some of the normal modes which appear in the spectra even of crystalline short-chain alkanes are space-group-symmetry-forbidden in the spectra of polyethylene. One must also keep in mind that either an isolated polymethylene chain or a real polyethylene crystal has centers of symmetry. This forbids the simultaneous appearance of any band in both the infrared and the Raman spectrum. Shorter chain alkanes, particularly in the liquid state, will be free of this restriction. Nevertheless, an examination of the vibrational spectra of polyethylene is instructive. The probable infrared spectrum, Raman spectrum and assignment of frequencies for a hypothetical, isolated extended polymethylene chain is given in Table 3.5. The CH stretching and CH_2

Table 3.5.　Vibrational Spectra of a Hypothetical,
Isolated, Polymethylene Chain

Raman	Infrared
2926 w	2924 s B_{3u} CH stretching
2890 s B_g CH stretching	2850 s B_{2u} CH stretching
2845 s A_g CH stretching	
2718 w	
1458 w	
1435 s A_g CH$_2$ bending	1463 s B_{3u} CH$_2$ bending
1412 w B_{2g} CH$_2$ wagging	1372 m B_{1u} CH$_2$ wagging
1293 s B_{3g} CH$_2$ twisting	
1177 w B_{1g} CH$_2$ rocking	720 s B_{2u} CH$_2$ rocking
1126 s B_{2g} skeletal	
1062 s A_g skeletal	
— A_u CH$_2$ twisting	

bending frequencies shown are characteristic of most saturated hydrocarbons. The infrared band near 720 cm^{-1}, generally absent in the Raman spectrum, is a good indication of an alkane chain, as is the Raman band near 1300 cm^{-1}. The 720 cm^{-1} band of the liquid generally splits into a doublet at 720 cm^{-1} and 730 cm^{-1} when a sample is crystallized. Raman bands of alkanes with more than 15 carbon atoms tend not to split when a liquid sample is crystallized. The Raman spectra of short chain alkanes have lattice bands below 300 cm^{-1}, which shift in a regular manner with chain length and which are characteristic of the particular alkane. Characteristic frequencies of normal alkanes are given in Table 3.6a. Many liquid alkanes

Table 3.6a.　Infrared and Raman Bands of Normal Alkanes

Vibration	IR	R
Carbon-hydrogen stretch	2980–2800 s	2980–2800 vs
Methylene rocking	720 ms	Absent
Methylene twisting and wagging	1400–1200 mw	1300 ms
Methyl asymmetrical bend	1465 ms	1465 s
Methyl symmetrical bend	1380 mw	Absent
Methyl rocking	1135 w	
Carbon-carbon stretch	Difficult to observe	1000 w
CC torsion and bend	Difficult to observe	335–235 m (dependent chain length)
No assignment	Absent	1075^a

^{a} Double band.

have a doublet in the Raman spectrum near 1075 cm^{-1}. Comparison with the polymethylene spectrum indicates that this is probably due to skeletal stretching motions.

Characteristic frequencies of branched chain hydrocarbons are given in Table 3.6b. The band positions are only approximate and bands arising

Table 3.6b. Infrared and Raman Bands of Branched Chain Alkanes

Assignment	IR	R
Tertiary CH stretch	Difficult to observe	Difficult to observe
Tertiary CH bend	Difficult to observe	1350 mw
Isopropyl symmetric bend	1380 m	
	1360 m	Weak or absent
Isopropyl CC stretch	955 w	955 w
Isopropyl deformation	1170 ms	1170 w
Isopropyl deformation	835–795 w	835–795 ms
Isopropyl deformation	Not observed	1345 mw
Tertiary butyl symmetrical bend	1400 ms[a]	Weak or absent
Tertiary butyl deformation	1250 ms	1250 ms
Tertiary butyl deformation	1205 ms	1205 ms
Tertiary butyl deformation	930 w	930 ms

[a] Double band.

from other sources may fall near the ones given. Nevertheless, in many cases, reasonable guesses may be made by using the table. A specific application is the identification of the nature of chain branching in polyethylene.[39] An examination of the infrared and Raman spectra of polypropylene, as well as the spectra of shorter chain hydrocarbons, makes it appear that tertiary methyl groups should have a very strong infrared absorption at 1382 cm^{-1} and a weak Raman band at this position. A weak infrared absorption and a strong Raman band should appear at 1333 cm^{-1}. Other characteristic bands are medium infrared and Raman bands, at 809, 975, and 1160 cm^{-1}. The infrared spectrum of polyethylene has no absorption at all at 974 cm^{-1}, showing that the side chains consist of ethyl, or longer, branches.

Some general observations may be made on the spectra of branched alkanes. Their spectra contain more bands than those of the normal alkanes. In the shorter-chain branched alkanes, the 1300 cm^{-1} Raman band may be absent. In longer chains, of course, this band will appear, from the linear portions of chain. If a compound contains aromatic rings as well as branched alkyl chains, it may be very difficult to detect the bands due to the latter. Branched chain alkanes often have a strong Raman band near 960 cm^{-1}, which is a good characteristic frequency if no aromatic rings are present. The absence of the latter can be demonstrated from the absence of characteristic aromatic ring frequencies.

ALKENES, ALKYNES, NITRILES, AND ISOCYANATES

The olefinic group frequencies are shown in Table 3.7. Olefines may be identified by very strong Raman bands near 3000, 1650, and 800–1400 cm^{-1} region. In the infrared spectrum, the 3000 and 1650 cm^{-1} bands are weak and very sharp. The out-of-plane infrared bending band between 850 and 1000 cm^{-1} is strong and is absent in the Raman spectrum. As was noted before, if the molecule has a center of symmetry, or if the double bond is in a symmetrical environment, a given olefinic band will not be strong in both the infrared and Raman spectrum.

Substitution of the double bond causes the intensities of the olefinic bands to increase. Substitution of the double bond with fluorine can cause major shifts in band positions and intensities.

Table 3.7. Olefinic Group Frequencies

Vibration	Infrared	Raman
A. Vinyl Group ($RCH=CH_2$)		
CH asymmetric stretch	3080 *m*	3080 *m*
CH symmetric stretch	3010 *m*	3010 *m*
CH_2 overtone	1820 *mw*	Absent
CC stretch	1640 *ms*	1640 *m*
CH_2 in-plane bend	1415 *m*	1415 *m*
CH in-plane bend	1300 *mw*	1300 *m*
CH out-of-plane bend	990 *ms*	Absent
CH_2 out-of-plane bend	910 *s*	Absent
Special Assignments		
1. $COCHCH_2$ out-of-plane	980 and 960	Absent
2. $NCCHCH_2$ out-of-plane	960 (one band only)	Absent
3. $ClCHCH_2$ out-of-plane	895 (one band only)	Absent
4. $ROCHCH_2$ out-of-plane	815	Absent
5. $RSCHCH_2$ out-of-plane	965 and 860	Absent
6. Deformation	—	435 *m*
B. Trans CHRCHR		
CH stretch	3025 *m*	3025 *m*
CC stretch	1675 *va*	1675 *m-s*
CH out-of-plane bend	970 *s*	Absent
Special Assignments		
1. RCHCHCN out-of-plane	955 *ms*	Absent
2. RCHCHCl out-of-plane	925 *ms*	Absent
3. ClCHCHCl out-of-plane	890 *ms*	490
4. Deformation	—	210

Table 3.7. (*Continued*)

C. Cis CHRCHR

CH stretch	3025 *m*	3025 *m*
CC stretch	1645 *m*	1645 *ms*
CH in-plane bend	1410 *m*	1410 *w*
CH out-of-plane bend	690 *ms*	Absent

Special Assignments

1. Unsaturated rings	Dependent on ring size	
2. Deformation	—	580
		415
		295

D. RRCCH$_2$ Groupings

CH stretch	3090 *m*	3090 *m*
CH overtone of bend	1785 *mw*	Absent
CC stretch	1650 *m*	1650 *m*
CH out-of-plane bend	890 *s*	Absent

Special Assignments

Deformation	—	435 *m*
		395 *m*
		260 *m*

The presence of electronegative groups raises the CC stretch and lowers the CH out-of-plane bend frequency. The opposite occurs for unsaturated groups.

E. RRCCHR Groupings

CH stretch	3030 *w*	3030 *m*
CC stretch	1680 *w*	1680 *m*
CH out-of-plane	815 *m*	Absent

F. RRCCRR

CC stretch	1680 *w*	1680 *m*

In conjugated dienes, trienes, and polyenes, there can be two or more double-bond stretching frequencies near 1600 cm^{-1}. It should be remarked that the CN double bond may also give rise to a stretching frequency near 1650 cm^{-1}.

Ketene-imine compounds are similar in structure to C=N compounds. They have a characteristic frequency near 2000 cm^{-1}.

The group frequencies of acetylenes and nitriles are given in Table 3.8. Their most characteristic frequency is a very strong Raman band in the 2150–2250 cm^{-1} region of the spectrum. This band will be sharp in the infrared. It is the only characteristic frequency for disubstituted acetylenes

Table 3.8. Group Frequencies of Acetylenic Compounds

Group	Infrared Bands	Raman Bands
Carbon-hydrogen stretch	3350 *m*	3350 *s*
Carbon-carbon triple bond stretch (in monoalkyls)	2125–2118 *mw*	2125–2118 *s*
Carbon-carbon triple bond stretch (in dialkyls two bands appear)	2305 *w*; 2227 *w*	2305 *s*; 2227 *s*
Carbon-carbon triple bond stretch (in conjugated aryl and olefinic)	2260–2090[a]	2260–2090[a]
Carbon-hydrogen bend	650–600 *s*	Absent
Carbon-hydrogen wag	700–625	Absent
Overtone of carbon-hydrogen bend	1400–1200 *m*	1400–1200 *s*
Nitriles	2500 *m*	2500 *s*

[a] Conjugation intensifies this band.

and nitriles. In monosubstituted acelytenes there is a strong CH stretching band near 3350 cm^{-1}. In asymetrically substituted acelytenes and in nitriles, the triple-bond stretching frequency may be fairly intense in the infrared.

Conjugation, cumulation or fluorine substitution increases the intensity and lowers the frequency of the $-C\equiv C-$ stretching band. Isocyanates and thiocyanates may have strong stretching frequencies near 2250 and 2050–2150 cm^{-1}, respectively. Thiocyanates may be identified by strong, broad Raman bands near 700 cm^{-1}.

AROMATIC COMPOUNDS

Aromatic compounds, like alkenes, have sharp infrared bands above 3050 cm^{-1} and two sharp, medium-to-strong infrared bands in the 1400–1650 cm^{-1} region. The Raman bands in the corresponding regions tend to be strong and polarized (Table 3.9). Generally only one Raman band appears near 1600 cm^{-1}. It is possible to confuse the spectrum of an alkene with that of an aromatic compound. The availability of both infrared and Raman spectra is particularly important here. It is desirable to have instrumentation to record both spectra on the same chart. This greatly facilitates examination of the spectra. (See "Worked Samples," p. 113). The infrared spectrum has a band pattern in the region 1700–2000 cm^{-1} which is absent in the Raman spectrum and is characteristic of the type of substitution (Fig. 3.1). This is a good indicator of the presence of an aromatic ring. Another good indicator is a very strong, polarized Raman band near 1000 cm^{-1}.

Table 3.9.   Group Frequencies of Aromatic Rings Having at Least One CH
Group

Vibration	Infrared	Raman
A. *All substitutions have the following group frequencies*		
CH stretch	3080–3030 *w*	3080–3030 *m*
Combinations and overtones	2000–1660 *w*	Absent
CC stretch	1650–1400 *m-s*[b]	1650–1400 *m-s*
CH in-plane bends	1400–1000 *va*	1400–1000 *va*
CC stretch	Absent	1000[a]
B. *Bands for identifying monosubstituted rings*		
CH out-of-plane	900–860 *m*	900–860 *m*
	770–730 *s*	700–730 *m*
	570–445 *m*	570–445 *m*
Ring deformation	710–690 *s*	Absent
C. *Bands for identifying 1,2-substituted rings*		
CH out-of-plane	770–735 *s*	770–735 *m*
D. *Bands used to identify 1,3-disubstituted rings*		
CH out-of-plane bend	810–750 *s*	810–750 *m*
	480–450 *m*	480–450 *m*
Ring deformation	725–680 *m*	Absent
E. *Bands used to identify 1,4-disubstituted rings*		
CH out-of-plane bend	860–800 *s*	860–800 *m*
	570–480 *m*	570–480 *m*
F. *Bands used to identify 1,2,3-trisubstituted rings*		
CH out-of-plane bend	810–750 *s*	810–750 *m*
	725–680 *m*	Absent
G. *Bands used to identify 1,2,4-trisubstituted rings*		
CH out-of-plane bend	900–860 *m*	900–860 *m*
	860–800 *s*	860–800 *m*
H. *Bands used to identify 1,3,5-trisubstituted rings*		
CH out-of-plane bend	900–860 *m*	900–860 *m*
Ring deformation	710–690 *s*	Absent
I. *Bands used to identify 1,2,3,4-tetrasubstituted rings*		
CH out-of-plane bend	900–860 *m*	900–860 *w*
J. *Bands used to identify 1,2,3,5-tetrasubstituted rings*		
CH out-of-plane bend	900–860 *m*	900–860 *w*
K. *Bands used to identify 1,2,4,5-tetrasubstituted rings*		
CH out-of-plane bend	900–860 *m*	900–860 *w*
L. *Bands used to identify 1,2,3,4,5-pentasubstituted rings*		
CH out-of-plane bend	900–860 *m*	900–860 *w*

[a] This band appears near 1000 for all substitutions except 1,2,3, where it is at 840.
[b] Intensity of these bands alternates in the infrared and Raman.

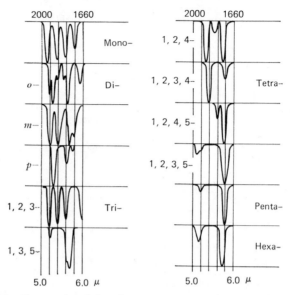

Fig. 3.1 Characteristic infrared overtone patterns of substituted benzenes.

Characteristic bands tend to appear in the 400–1000 cm^{-1} region. An experienced spectroscopist can generally use these both to confirm the presence of an aromatic ring and to identify the type of substitution.

Heterocyclic aromatic compounds are, structurally, quite close to benzene. Like benzene, they have a CH stretching frequency near 3020 cm^{-1} and a ring frequency near 1600 cm^{-1}. Pyridines and quinolines have strong infrared absorptions near 700, 1050, and 1200 cm^{-1}. In pyrimidines the high wavenumber ring frequency falls at 1520–1580 cm^{-1} in the infrared and there are infrared absorptions near 800 and 1000 cm^{-1}.

The pyrimidines and purenes are of great biological interest. Raman spectra of these materials have been published only recently.[40] These have not been analyzed in detail for stable group frequencies. One might note the presence of strong Raman bands near 3050 cm^{-1} and between 1500 and 1650 cm^{-1} and characteristic ring frequencies in the 700–1400 cm^{-1} regions, for adenine, uracil, cytosine and guanine rings. It should be mentioned that Lord and Thomas give infrared as well as Raman spectra and that the Raman spectra in solution change markedly as the pH is changed.

The polynuclear aromatics are recognizable as aromatics from their spectra, but the spectra are, in general, too complex for deduction of structural details. One can say, as a general rule, that the spectra will contain a large number of strong, sharp bands. A strong Raman band near 750 cm^{-1} and a strong infrared band near 950 cm^{-1}, may be identifiable.

CARBONYL COMPOUNDS, CARBOXYL COMPOUNDS, ESTERS, AND AMIDES

Carbonyl compounds are easily identified by one or more very strong, generally broad infrared absorption bands between 1600 and 1800 cm^{-1}. These bands are somewhat less strong in the Raman effect and may appear at different positions from the infrared bands if more than one carbonyl group is present. The contours and positions of the $-C=O$ stretching bands are often a good indicator of the type of carbonyl present. Other parts of the spectrum may prove helpful. Aldehydes, for example, generally have a weak infrared band near 2700 cm^{-1}. A saturated, open-chain ketone will generally have a band between 1700 and 1725 cm^{-1}. α-halogen or α,α'-dihalogen substitution will raise these values to 1725–1765 cm^{-1}. Four- and five-membered ring ketones have their carbonyl stretching frequency in the same region. On the other hand, aryl and diaryl, ketones, α,β-unsaturated ketones, β-diketones and quinones, have this band in the 1550–1700 cm^{-1} region. Ketones which can have *s-cis* and *s-trans* forms will have multiple, very strong bands in the 1600–1700 cm^{-1} region, in both the infrared and the Raman spectra.

Carboxylic acids have strong, multiple infrared and Raman bands in the 1600–1800 cm^{-1} region. The bands will tend to coincide in the two spectra, but the intensity patterns will be different. There is, generally, a strong broad, complex infrared band which stretches from about 2500 to 3000 cm^{-1}. This band is much weaker in the Raman spectrum and, at times, may be barely visible. There is, sometimes, a strong Raman band near 800 cm^{-1}. If a solution of the acid is alkalized and such a band disappears, this is good confirmation of the presence of a carboxylic acid.

Esters may be identified by one or two strong bands in the 1720–1800 cm^{-1} region and by a pair of medium to strong bands, one in the 1000–1100 cm^{-1} region and one in the 1100–1300 cm^{-1} region. The bands near 1700 cm^{-1} and in the 1100–1300 cm^{-1} region will generally be strong in the infrared spectrum, that near 1050 cm^{-1} will be strong in the Raman spectrum. As is usually the case, fluorination shifts all of these values to higher frequencies.

Anhydrides have two strong carbonyl frequencies, one in the 1700–1800 cm^{-1} region and one in the 1750–1880 cm^{-1} region. They also have a pair of bands of which one lies near 1000 cm^{-1} and the other at 1200–1300 cm^{-1}. There is some dependence of band position on the type of anhydride.

Lactones have a strong, broad infrared band around 1800 cm^{-1} and a strong, narrower Raman band at the same position. Often, two bands are observed. There may be other strong bands in the spectra, but they are of little diagnostic value.

The group frequencies of amides are listed in Table 3.10. Primary and secondary amides are rather easy to identify. The NH stretching, amide I

Table 3.10. Group Frequencies of Amides

Vibration	Infrared	Raman
A. Primary amides		
Asymmetric NH stretch	3500 (free)	3500 (free)
	3350 (bonded)	3350 (bonded)
Symmetric NH stretch	3400 (free)	3400 (free)
	3180 (bonded)	3180 (bonded)
Amide I (carbonyl stretch)	1680 (free)	Band is weaker in the Raman than in the infrared
Amide II (NH bend)	1620–1590 (free)	Band is weak in the Raman
	1650–1620 (bonded)	
CN stretch	1410 m	May be strong in the Raman
NH wag	750–600 mb	Band is weak in the Raman
B. Secondary amides		
NH stretch	3460–3400 (free)	Same as in the infrared
	3320–3270 (bonded in trans– form)	
	3180–3140 (bonded in cis– form)	Same as in the infrared
Amide I (carbonyl stretch)	1700–1670 (free)	Band is weak in the Raman
	1680–1630 (bonded)	
Amide II (NH bend)	1550–1510 (free)	Band does not appear in the Raman
	1570–1515 (bonded)	
Amide III (CN stretch)	1290 m	May be strong in the Raman
Amide IV (NH deformation)	620 w	Same as in the infrared
Amide V (NH deformation)	720 (bonded state only)	Same as in the infrared
Amide VI (NH deformation)	600 w	Same as in the infrared
C. Tertiary amides		
Amide I (carbonyl stretch)	1650 s	1650 m
D. Cyclic Amides		
Amide I for large rings	1680 (free)	Same as in the infrared
Amide I for Lactams	1750–1700 (free)	Same as in the infrared
	1760–1730 (free)	Same as in the infrared

and amide II bands are generally used to follow the conformational changes of proteins. The three bands form a quite distinctive pattern in the infrared spectrum. In the Raman spectrum, the NH stretching bands and the amide II band are quite weak. The amide I band retains moderate strength.

Tertiary amides have only the amide I band and may be difficult to identify. The bands labeled amide III-IV are sometimes useful for diagnostic purposes. Unfortunately, they are hard to pick out of complex spectra.

ALCOHOLS, AMINES, AND THIOLS

The most outstanding feature in the infrared spectrum of alcohols is a very strong band or series of bands from 3200–3600 cm^{-1}. The exact positions are dependent on the structure and degree of hydrogen bonding of the alcohol. Efforts should be made to examine an anhydrous sample, since water has a very strong absorption in this region of the spectrum. Other

Table 3.11. Group Frequencies for Alcohols and Thiols

Frequency	Infrared	Raman
OH Stretch		
A. Primary alcohols	3400–3230 *mb* (associated)	Weak or absent
	3670–3580 *msp* (dissociated)	Weak or absent
B. Secondary alcohols	3400–3200 *mb* (associated)	Weak or absent
	3660–3570 *msp* (dissociated)	Weak or absent
C. Tertiary alcohols	3400–3200 *mb* (associated)	Weak or absent
	3650–3560 *msp* (dissociated)	Weak or absent
D. 1, 2 diols	3600–3500 *msp*	Weak or absent
E. Chelates	3200–2500 *mb*	Weak or absent
F. Tropolones	3200–3000 *mb*	Weak or absent
OH In-plane bend		
A. Primary alcohols	1420–1380 *mb*	Weak or absent
B. Secondary alcohols	1420–1380 *mb*	Weak or absent
C. Tertiary alcohols	1410–1320 *mb*	Weak or absent
CO Stretch		
A. Primary alcohols	1075–1000 *mb*	Weak or absent
B. Secondary alcohols	1125–1090 *mb*	Weak or absent
C. Tertiary alcohols	1210–1000 *msb*	Weak or absent
D. ArCOH	1040–980 *msb*	Weak or absent
E. Cyclohexanols	1035–970 *msb* (axial)	Weak or absent
	1065–1035 *msb* (equitorial)	Weak or absent
OH Out-of-plane		
Primary, secondary,		
and tertiary alcohols	750–225 *ms*	Weak or absent
CH wag of alcohols	1330–1200 *m*	Medium intensity in the Raman
CC stretch	950–850 *vw*	Strong intensity in the Raman
SH stretch	Weak	2550–2600 *s*
CS stretch	Weak	600–700 *s*
PH stretch	2350–2440 *m*	2350–2440 *m*

bands may be found in the regions 1300–1500 cm^{-1} (in-plane bending), 1000–1300 cm^{-1} (CO stretching) and 650 cm^{-1} (out-of-plane bending). These bands all tend to be weak in the Raman spectrum.

As may be seen from Table 3.11, once it is established that one is dealing with an alcohol, something may be learned from the spectra about what kind of an alcohol is involved. Solution with non-polar solvents and temperature changes will markedly influence the positions and contours of OH stretching bands. This is a good means of identifying an alcohol.

Table 3.12. Infrared Group Frequencies of Amines

Group	Asymmetric NH Stretch	Symmetric NH Stretch	NH Bend	CN Stretch	NH Wag
RCH$_2$NH$_2$	3350–3330 ms	3450–3250 ms	1650–1590 s	1090–1070 m	850–750 sb
R$_3$CNH$_2$	3550–3330 ms	3450–3250 ms	1650–1590 s	1240–1170 s	850–750 sb
				1038–1020 s	850–750 sb
ArNH$_2$	3550–3330 ms	3450–3250 ms	1650–1590 s	1330–1260 s	850–750 sb
ArCHNH$_2$	3550–3330 ms	3450–3250 ms	1650–1590 s	1140–1080 s	850–750 sb
				1045–1035 w	850–750 sb
Ar$_3$CNH	3550–3330 ms	3450–3250 ms	1650–1590 s	1045–1035 w	850–750 sb
RCH$_2$NHR	3500–3300 w	—	?	1145–1130 ms	750–700 s
RCH$_2$NHAr	3500–3300 w	—	1510 ms		
ArCH$_2$NHAr	3500–3300 w	—	1510 m	1340–1320 s	750–700 s
				1315–1250 s	
ArNHR	3500–3300 w	—	?	1340–1320 s	750–700 s
				1315–1250 s	
R$_2$CHNHR	3500–3300 w	—	?	1190–1170 s	750–700 s
Ar—N=N—NHAr	3500–3300 w	—	1522 m	1178 m	—
RNR$_2$	—	—	—	—	
ArNR$_2$	—	—	—	1380–1265 s	—

Amines have their NH stretching frequencies in the same region of the spectrum as the OH stretching frequencies, but the NH bending band tends to fall higher, at 1500–1650 cm^{-1} (Table 3.12). The Raman bands of amines arising from NH motions tend to be weak, although the NH stretching band may be strong. If there is extensive hydrogen bonding, then the NH stretching band may be weak and broad in the Raman spectrum. Amine bands are less affected by hydrogen bonds than are OH bands.

Thiols are easily identified by strong Raman bands at 2550–2600 and 600–700 cm^{-1}. These are weak in the infrared spectrum.

ETHERS

The COC group in noncyclic ethers has a symmetric and an asymmetric stretch. The former will tend to be strong in the Raman spectrum, the latter in the infrared spectrum. Table 3.13 lists characteristic frequencies for several types of ethers. It is generally not difficult to identify ethers, unless the ether group is only a small part of a large molecule.

Table 3.13. Group Frequencies of Ethers

	Infrared	Raman
Alkyl	1060–1150 vs	1060–1150 w
	950 w	950 s
	300– 600 w	300– 600 s
Aryl ethers	1230–1270 vs	—
Expoxy compounds	1270 s	1270 vs
	870 s	870 m-s
Cyclic ethers	1070–1140 vs	1070–1140 m
	800– 950 m	800– 950 vs

NITRO COMPOUNDS, SULFUR-OXYGEN, AND PHOSPHOROUS-OXYGEN COMPOUNDS

The characteristic group frequencies of the nitro group are relatively independent of the rest of the molecule, which makes this group easy to identify. The symmetric stretching band at 1350–1400 cm^{-1} is strong in the infrared spectrum, but only of medium strength and depolarized in the Raman spectrum. The NO_2 bending band at 650–700 cm^{-1} is of medium intensity in both spectra, but should be polarized in the Raman spectrum. A band at 800–900 cm^{-1} due to the $C-(NO_2)$ stretching motion, is strong in the Raman spectrum and weak in the infrared spectrum. It should be polarized in the Raman spectrum. In aromatic compounds, the NO_2 stretching bands are particularly intense in the Raman spectrum.

There is a variety of sulfur-oxygen compounds, which give rise to strong infrared absorptions. The corresponding Raman bands also tend to be strong, but, unfortunately, few studies of group frequencies have been made. Of those groups having two strong characteristic infrared bands, the lower should be strong and polarized in the Raman spectrum.

Phosphorous-oxygen compounds tend to have a single, strong infrared and Raman band.

Some correlations are shown in Table 3.14.

Table 3.14. Infrared Group Frequencies of Nitro Compounds, Sulfur-Oxygen Compounds, and Phosphorous-Oxygen Compounds[a]

Compound	Physical State	NO_2 Asymmetric Stretch	NO_2 Symmetric Stretch	CN Stretch	NO_2 Bend
CH_3NO_2	L	1558 vs, mb	1401⎫ 1379⎭ ms, sp, db	918 mb	657 m
RCH_2NO_2	L	1556–1545 vs, mb	1388–1368 ms, sp	870–820 mb	
$RC(CH_3)_2NO_2$	L	1553–1530 vs, mb	1359–1342 ms		
$RCH(CH_3)NO_2$	L	1549–1545 vs, mb	1364–1357 ms, sp	870–820 mb	
$RCH(X)NO_2$ (X = Cl, Br)	L	1580–1556 vs	1368–1340 ms	860–820 mb	
RCX_2NO_2	L	1597–1569 vsb	1339–1323 ms	860–820 mb	
CCl_3NO_2	L	1610 vsb	1307 ms	860–820 mb	
CF_3NO_2	L	1620 vsb	1315 ms	863 mb	604 m
UNO_2	L	1550–1500 vs, mb	1360–1290 ms		
$C_6H_5NO_2$	L	1583 vsb	1351 ms	852 mb	677 m
$o-HOC_6H_4NO_2$	L	1536 vsb	1315 ms	820 mb	655 m
$p-NH_2C_6H_4NO_2$	KBr	1481–1470 vsb	1335⎫ 1302⎭ msb	857 mb	745 m
5-Nitro-2 amino-pyridines			1310–1270 msb		
3-Nitro-2 amino-pyridines			1250–1210 msb		
$R-O-NO_2$		1660–1620 sb	1285–1270 msb		
$Na^+CH_2NO_2^-$	Sl	1277 vs	1033⎫ 1018⎭ vs		689⎫ 677⎭ s
$Na^+(CH_3)_2$ CNO_2^-	Sl	1176 sh 1163 vs	944 s		624 m
$N^{14}O_2$	vp	1618 s	1318 s		
$N^{15}O_2$	vp	1580 s	1305 s		
NO_2^+	vp	1275–1235 sb	835–820 msp		
$-S=O$		1040–1060 s			
$-SO-OH$		1090 s			
$R-SO_2-R$		1140–1160 s and 1300–1350 s			
$-O-SO_2-O$		1150–1230 s and 1350–1440 s			
SO_3H		1030–1060s and 1150–1210 s			
$P=O$		1200–1350 s (strong in Raman)			
$-P-O-C$		1000–1250 m-s			

[a] L = liquid; Sl = Solid; vp = vapor.

CARBON-HALOGEN AND METALORGANIC COMPOUNDS

Carbon-halogen compounds have all of the characteristic frequencies associated with the CX, CX_2 or CX_3 group. With the exception of CF compounds, the characteristic frequencies lie below 800 cm^{-1}. The bands associated with these groups lead to strong infrared absorption and Raman scattering. One of the CX stretching bands will be highly polarized. The positions of the associated bands can often be located by analogy to the substituted methanes, ethanes and 2,2-substituted propanes. Similar remarks, as was mentioned earlier, are valid for the metalorganic compounds of silicon, lead, mercury and similar metals. As an example, $(C_2H_5)\ GeCl_3$ has strong Raman bands at 136, 174, 397 and 425 cm^{-1}, corresponding to Raman bands at 132, 171, 397 and 451 cm^{-1} of $GeCl_4$. A Raman band at 596 cm^{-1} may be compared to the 595 cm^{-1} GeC stretching band of $Ge(CH_3)_4$.

There are several difficulties in interpreting the spectra of a metalorganic compound, if the structure is not known. One is that the group frequencies are crowded into the region of the spectrum below 600 cm^{-1}. The different metal-carbon frequencies do not differ much from one another. There is also a tendency for frequencies from different normal modes to "pile up." The two CPbC bending bands of $Pb(CH_3)_4$, for example, appear as a single, broad Raman band around 130 cm^{-1}. Nevertheless, if an elemental analysis is available, a good guess as to structure can often be made. It is advantageous to examine the Raman spectra of carbon-halogen and metalorganic compounds in the crystalline state. The spectra of these compounds frequently exhibit effects due to conformational isomerism. This is frozen out in the crystal.

Some group frequencies are shown in Table 3.15.

Table 3.15. Group Stretching Frequencies of Some Carbon–Halogen and Metal–Carbon Bonds

C-F	1000–1400[a]
C-Cl	600–800
C-Br	500–600
C-I	500
Phenyl-Cl	750
Si—C	1250
Ge—C	550–600
Sn—C	520
Pb—C	460
Hg—C	550

[a] May be weak in Raman.

INORGANIC COMPOUNDS

The possible number of inorganic compounds is so vast that little can be deduced from the spectrum of an unknown, without some prior knowledge. Some special classes of compounds give a hint as to their nature from the fact that they are extremely powerful Raman scatterers. If an unknown compound has one or more strong Raman bands in the region 1900–2100 cm^{-1} and is an overall powerful Raman scatterer, one would suspect a metal carbonyl. Ions, like nitrates, sulfates, thiocyanates, etc., have characteristic frequencies which are relatively independent of the nature of the counter-ion or of the physical state. These ions, being, in essence, small symmetrical molecules, will have a number of Raman bands predictable from their geometry. The polarization of these bands is also predictable from the geometry. Some illustrative examples are given in Table 3.16.

Table 3.16. Group Frequencies of Ions, Mean Values, cm^{-1}

$NO_3{}^-$	1400	1050 p^a	820	720
$BO_3{}^-$	1450	910 p	700	
$CO_3{}^-$	1415	1060 p	880	680
$NO_2{}^-$	1330	815 p	590 p	
$(OCN)^-$	2192 p	857 p	616	
$(CSN)^-$	2026 p	750 p	500	
$(ClO_3)^-$	980	930 p	625 p	480
$(BrO_3)^-$	830	800 p	440 p	350
$(IO_3)^-$	810	780 p	360 p	330
$NH_4{}^+$	3145	3040 p	1680	1400
$ND_4{}^+$	2350	2215 p	1215	1065
$PO_4{}^{3-}$	1080	935 p	550	420
$SO_2{}^{2-}$	1100	980 p	615	450
$ClO_4{}^-$	1060	930 p	625	460
$IO_4{}^-$	840	795 p	340	280
$N_3{}^-$	2070	1350	635	
OH^-	3300–3600 p			

[a] The p deonotes a highly polarized Raman band.

Where possible, the Raman spectrum of a sample suspected of containing an ion should be examined in a not-too-concentrated aqueous solution. The bands listed will often split in the crystal. The infrared spectrum is less useful for identifying an ion, since the infrared absorption bands tend to be broad and complex.

Various metal oxides and minerals, such as the garnets, have quite complex Raman spectra. These are useful for identifying a mineral, or for distinguishing between polymorphic forms of the same mineral, if a library of spectra is available.

Crystalline materials, whether organic or inorganic, will generally have lattice bands below 300 cm^{-1}.

SOME WORKED SAMPLES

1. $F_3C-C\equiv C-C\equiv C-CF_3$. In the spectra of Figure 3.2 we note first, the absence of any but very weak bands between 1300 and 2000 cm^{-1} and of any bands above 2300 cm^{-1} (spectra not shown). This eliminates any compound containing CH groups, NH groups, carbonyl groups, $-C=C-$, etc. With the possible exception of the band near 1180 cm^{-1},

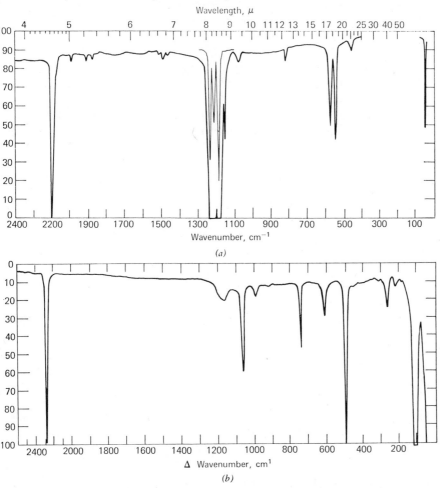

Fig. 3.2 The infrared and Raman spectra of $F_3C-C\equiv C-C\equiv CF_3$. (Courtesy of D. H. Lemmon.) (a) Infrared spectrum (vapor), 10-cm path. A 45-cm^{-1} band at 120 torr; rest of spectrum at 10 torr except the upper trace of 1200 region which was ~0.1 torr. (b) Raman spectrum (liquid), 80 mW 6328 Å excitation. Slits 6.7 cm^{-1} × 1 mm, photon count 5k, time constant 1 sec, scan 200 cm^{-1}/min, cooled FW 130 detector.

no Raman band appears in the infrared spectrum and vice versa. This indicates that we are probably dealing with a relatively simple, centrosymmetric molecule. There is a strong, sharp infrared absorption at 2200 cm^{-1} and a strong Raman band at 2240 cm^{-1}. These indicate triple bonding. Since there are two bands, there are at least two, conjugated triple bonds. The very strong, multiple infrared absorption band near 1200 cm^{-1}, is indicative of a fluorocarbon. Ether or ester bands fall here, but the absence of a strong Raman band near 940 cm^{-1} and the absence of CH bands makes the latter unlikely. The complex nature of the infrared band at 1200 cm^{-1} indicates that the fluorine is probably in the form CF_2 or CF_3. Possible structures are $F_3C-C{\equiv}C-C{\equiv}N$ or $F_3C-C{\equiv}C-C{\equiv}CF_3$. If nitrogen is known to be absent, the latter is a likely structure. If an elemental analysis is available, this structure is certain.

2. Mesitylene (1,3,5-Trimethyl Benzene). The infrared and Raman spectra of this compound are shown in Figure 3.3. We first note the group

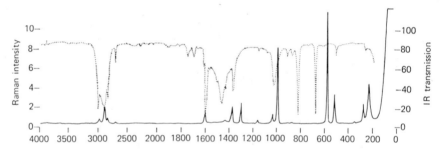

Fig. 3.3   The infrared and Raman spectra of mesitylene. Dashed line: infrared spectrum; solid line; Raman spectrum. (Courtesy of the Perkin-Elmer Corporation.)

of bands near 2900 and 1450 cm^{-1}, indicating aliphatic CH and the bands at 1600 and 3000 cm^{-1}, characteristic of olefines and aromatic compounds. The two infrared bands at 1750 and 1820 cm^{-1} and the very strong Raman band at 1000 cm^{-1} indicate an aromatic compound. The relatively simple spectrum favors a phenyl compound rather than a polynuclear aromatic. (The weakness of the 1600 cm^{-1} and 3000 cm^{-1} Raman bands is deceptive. This is probably an artifact due to the falloff of the photomultiplier tube sensitivity at longer wavelengths.)

The infrared bands at 670 and 820 cm^{-1} are consistent with 1, 2, 3, or 1, 3, 5 trisubstitution, although the 820 cm^{-1} band lies a little outside the normal range for either. There is a number of coincident bands, which eliminates any type of substitution leaving a center of symmetry. The pattern of the infrared bands at 1750 and 1920 cm^{-1} would lead one to favor 1, 3, 5 substitution over 1, 2, 3 substitution, although, again, the pattern is somewhat atypical.

One more important conclusion may be drawn from the spectra. It will be noted that some bands appear in the infrared spectrum and not in the Raman spectrum, some in the Raman spectrum and not in the infrared spectrum, and some in both spectra. This is an indication that the missing bands are symmetry-forbidden, but that no center of symmetry is present. This explains why the 820 cm^{-1} band, normally present in the Raman spectrum of 1, 3, 5 substituted aromatic compounds, is absent.

3. Indene.   The spectra of this polynuclear aromatic appear in Figure 3.4. The infrared and Raman bands at 1600 and above 3000 cm^{-1} denote

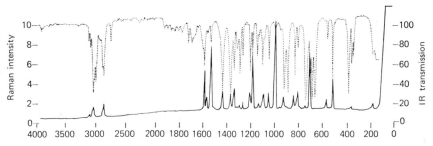

Fig. 3.4   The infrared and Raman spectra of indene. Dashed line: infrared spectrum; solid line: Raman spectrum. (Courtesy of the Perkin-Elmer Corporation.)

unsaturation. The patterns of infrared and Raman bands between 1800 and 2000 cm^{-1} show that the compound is aromatic. The large number of strong, sharp bands is characteristic of a polynuclear hydrocarbon. In spite of the fact that the 1450 cm^{-1} band is not unambiguiously assignable to a CH_2 bending motion, combined with the 2950 cm^{-1} band, it implies the presence of aliphatic CH. Lastly, the large number of frequencies which appears in both the infrared spectrum and the Raman spectrum implies that the symmetry of the molecule is low.

This is as much as can reasonably be concluded from the complicated spectra of this molecule.

4. Dibutyltin Dichloride.[41]   The Raman spectrum shown in Figure 3.5a was excited with a mercury arc, photographed and read with a densitometer. This example was chosen because it is one of the few extant studies of a metalorganic compound where actual spectra were shown. One notes, first, the absence of bands above 3000 or above 1450 cm^{-1}. This automatically eliminates any unsaturated hydrocarbon, nitrile, etc. The spectra are complex enough to arise from a polynuclear aromatic, but the breadth of

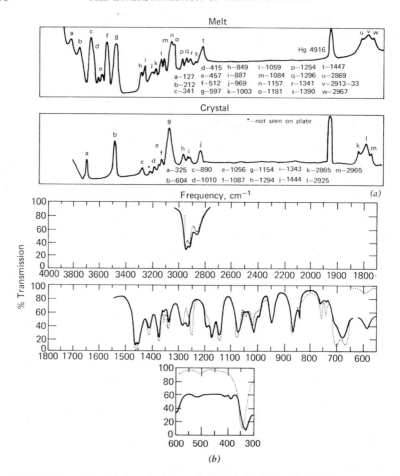

Fig. 3.5 The Raman and infrared spectra of dibutyl tin dichloride. (*a*) Raman spectrum. *Top*, melt; *bottom*, crystal. (*b*) Infrared spectrum. Solid line: crystal; dashed line: melt. (CS_2 solution below 600 cm^{-2}).

the bands and the absence of the characteristic aromatic bands eliminates this possibility. We know, then, that we are dealing with some kind of saturated hydrocarbon. The group of very strong Raman bands lying below 600 cm^{-1} makes one think immediately of a carbon-halogen compound or of a metalorganic compound. While the absolute scattering intensities of the low-lying Raman bands are not readable from the graph, on the original photographic plate, these bands, as well as the Rayleigh band were extremely strong. This favors a metalorganic compound. This disappearance of the Raman band at 512 cm^{-1} and the weakening of the infrared band at

710 cm^{-1} in going from melt to crystal indicates that several configurational isomers were present. We thus have a metalorganic compound substituted with propyl or longer chains. It would be difficult to go further without more knowledge. The 597 cm^{-1} band, for instance, arises from a SnC stretching motion. Just looking at the spectra, with no other information, it would be just as reasonable to assign this band to a CCl or CBr group.

4

SOME RECENT APPLICATIONS OF LASER RAMAN SPECTROSCOPY

4.1 INTRODUCTION

This chapter describes some particularly interesting applications of laser Raman spectroscopy to problems in chemistry. An encyclopedic coverage is not given.* Instead, particularly interesting applications and approaches have been selected for illustrative purposes. It is to be regretted that nothing seems to have been done with the application of laser Raman spectroscopy to quantitative chemical analysis.

4.2 POLYMERS AND BIOLOGICAL MATERIALS

With a few exceptions, all of the published Raman spectra of polymers, including biological macromolecules, have appeared since the advent of laser Raman spectroscopy. There are particular difficulties associated with the examination of these materials. They are often turbid and invariably fluoresce, even when red excitation is used. The fluorescence tends to be more severe in solution than in solid specimens. It is particularly strong in alkaline aqueous solutions. The considerations of signal-to-noise ratios, in the presence of a continuous background, therefore apply with particular force to these materials (Section 2.3). The source of this fluorescence is still unknown. It may be due to small traces of impurities or to absorption in the long-wavelength wing of the lowest electronic transition. It is often impossible to remove.

Analysis of the Raman spectra of polymers has some features in common with analysis of the Raman spectra of crystals. In polymers containing only one kind of monomer, the spectrum mirrors primarily the spectrum of a single repeat unit. This holds quite rigorously if the polymer can be orientated by stretching. If this is done, there is a preferred axis of symmetry. The polarizability tensors will then bear a resemblance to those of uniaxial crystals. By changing the orientation of the specimen relative to that of the

---

* The reader is referred to the excellent articles which appear biennially in *Analytical Chemistry*.

incident beam, deductions concerning the origin of observed bands may be made.[42,43,44] In the solid, amorphous state, polymers without cross-linking or very bulky side groups will tend to assume a random flight configuration. This is also the common configuration in solution.[45] Ionized polyelectrolytes will tend to be in the extended configuration. The change is often reflected in the Raman spectra of the solutions.[46] Crystalline polymers may appear in a number of configurations, depending on whether or not the chains are in an extended or in a helical configuration and depending on whether or not the polymer is isotactic, syndiotactic, etc. Again, major differences may appear between the Raman spectra of the various forms.

Biological macromolecules form a special class of polymer. In general, a great many different kinds of monomer units go into making up the chain. The individual molecules particularly in the case of proteins, are often internally cross-linked. These features make the spectra exceptionally difficult to interpret.

It is only for the simple polymers, like teflon or polyethylene, that a complete assignment of observed frequencies to normal modes can be made. Most often, an analysis is based on group frequencies and may be aimed at detecting changes in chain configuration with environment in solution. Cain and Harvey[47] give Raman spectra of a group of commercial polymers. Of particular interest are strong bands at 510 and 650 cm^{-1} due to S—S and C—S linkages, in a polysulfide rubber. They find C=C linkages in polybutadiene to produce strong Raman bands, but weak infrared absorption bands. The carbonyl groups of polycarbonates and polyesters, by contrast, give rise to strong infrared absorption, but weak Raman scattering. The Raman spectrum of polyethylene is reported by Freznel, Bradley, and Mathur[48] and by Snyder.[49] The former work is of interest in that the workers not only confirm the previously reported Raman bands due to polyethylene chain vibrations, but report some bands due to interchain lattice vibrations of the crystal. The observed frequencies agree well with those predicted from neutron scattering experiments.

The Raman spectrum of polypropylene, the prototype branched hydrocarbon, is treated by Schaufle[50] and by Zerbi and Hendra[51] (see Figure 1.2b). These workers give spectra far superior to those of the pre-laser Raman literature. They note marked differences between the spectra of the isotactic and the syndiotactic forms.

Boerio and Koenig[52] present an in-depth experimental and theoretical analysis of the Raman spectrum of teflon, including measurements at −50 and −130°C and measurements of low-molecular-weight homologs of teflon. The results were used to calculate the dispersion curves for a teflon helix and to assign the observed Raman bands to the normal modes of vibration. Cornell and Koenig[42] analyse the Raman spectra of orientated samples of polystyrene. An assignment of frequencies to the normal modes was facilitated by study of the change of depolarization ratio with orientation.

Tobin[46] gives Raman spectra of polymethacrylic acid as the amorphous solid, in aqueous solution and in alkaline aqueous solution. The spectra of the solid and of the neutral aqueous solution are virtually identical, implying that the chain configurations are similar in both cases. There are changes, including the disappearance of a strong OH bending band at 759 cm^{-1} when the aqueous solution is made alkaline.

The Raman spectrum of polyoxymethylene is presented by Sugeta, Miyazawa and Kajiura[53] but only a limited analysis is made.

Although interest in the Raman spectra of biological macromolecules is intense, only a very few spectra have appeared so far. The spectra of crystalline lysozyme, pepsin and α-chymotrypsin are given in a note.[54] The Raman spectra of calf thymus and of salmon testes DNA have been analyzed in some detail.[55] The infrared spectrum has an absorption band at 1690 cm^{-1}, associated with the ordered B-helix. No Raman band ever appears at this position for calf thymus DNA. While it does for salmon testes DNA, the band intensity is insensitive to the degree of order. Alkalization of 10% aqueous gels does not seem to unzip the double helices. Several assignments of observed Raman bands are made. The Raman spectrum of solid salmon testes DNA is shown in Figure 2.4.

Fanconi, Tomlinson, Nafie, Small and Peticolas[44] treat theoretically the Raman scattering expected from oriented helices. They then apply their theory to Raman spectra of fibers of poly-1-alanine and assign many of the frequencies to the chain normal modes. They also report spectra of poly-adenylic, polyriboguanilic, polyuridylic and polycytidylic acid in solution. Again, they make a partial assignment of frequencies to the normal modes. Tomlinson and Peticolas[56] have recently demonstrated the existence of Raman hypochromism in polyadenylic acid. When the double helix uncoils with increasing temperature, the 720 cm^{-1} Raman band loses intensity. This effect should prove an extremely valuable adjunct to ultraviolet hypochromism, since the Raman bands may be assigned to specific chemical groupings. The foregoing review requires some general comments. The first is that almost none of the spectra reported could have been recorded at all with pre-laser Raman equipment. The second is that some of the results are beyond the power of infrared absorption spectroscopy to obtain. The C—S and S—S bands in the polysulfide rubber probably could not have been picked out in an infrared spectrum. The studies of biological macromolecules and of polymethacrylic acid in aqueous solution simply could not have been made using infrared absorption spectroscopy. On the other hand, a thorough treatment of the vibrational spectra of polymers requires measurement of the spectra of oriented samples using polarized radiation. For work on polymers, both the infrared absorption spectrum and the Raman spectrum are really essential.

4.3 MINERALS AND INORGANIC GLASSES

A very lengthy literature exists on the Raman spectra of minerals and glasses.[57] Unfortunately, many of the spectra are of poor quality, so that the conclusions drawn from them are of limited value. Recently, laser Raman spectroscopy has been applied to the study of these materials. Such studies have proved to be very useful. Often, glasses and minerals are extremely powerful absorbers of infrared radiation, may be available only as refractory powders and may have infrared absorption spectra which are much more complex than the Raman spectra. While the more complex spectrum may contain more information, the information will be easier to extract if a simpler spectrum has first been analysed.

Griffiths[58] gives the Raman spectra of some twenty-five minerals containing MO_3 and MO_4 functional groups, where M is a metal (carbon, silicon, phosphorous, arsenic, vanadium, sulfur, chromium, molybdenum, or tungsten.) The minerals were examined in the form of 5–10 mg of a fine powder. The spectra were found to arise primarily from the vibrations of the functional group ($(PO_4)^{3-}$, etc.) Only some very low-frequency bands seemed to be associated with vibrations of the crystal lattice. The method makes it possible to distinguish between minerals with the same M.

Beattie and Gilson[59] give a lengthy theoretical and experimental study of a number of inorganic single crystals, including the minerals rutile, cassiterite and anatase.

By using properly oriented single crystals, they measured elements α_{ij} of the polarizability tensor, for the normal modes of a number of materials.

The analysis of the Raman spectra of minerals and similar inorganic crystals is greatly facilitated by theoretical analysis, mostly based on symmetry considerations. This aid is not available for the interpretation of the spectra of inorganic glasses, which, in effect, consist of giant, disordered polymer molecules. Two laser Raman studies have appeared, one of a series of silica glasses[60] and one of B_2O_3 and of a series of $B_2O_3-Bi_2O_3-Al_2O_3$ glasses.[61] It was found, surprisingly, that the spectra of the silica glasses are relatively insensitive to large changes in composition. Crystallization, either to a micro-crystalline glass or to crystal quartz, did effect major changes in the spectra. Changes in local order could be detected by changes in the depolarization ratios of the Raman bands. A normal coordinate analysis of fused silica[62] yields a dispersion curve strongly resembling the observed Raman spectrum. The low wavenumber region shows marked temperature effects.[63]

Above 4% by weight of Bi_2O_3, the Raman spectrum of B_2O_3—Bi_2O_3—Al_2O_3 glasses change drastically from that of B_2O_3. A further increase in the Bi_2O_3 concentration serves mainly to change the relative intensities of two bands near 70 and 130 cm^{-1}.

4.4   INORGANIC SINGLE CRYSTALS

Laser Raman spectroscopy has found very wide application in the study of single crystals. As we explained in Section 2.10, the individual elements of the polarizability tensor for the normal vibrations of single crystals may be measured in the laser Raman spectra. Sometimes, the data are simply used to facilitate an assignment of frequencies to the normal modes, or to estimate the strength of intermolecular forces. There is a whole group of phenomena of interest to solid-state physicists, such as spin-wave scattering, polariton scattering and scattering from lattice defects, which may be observed with Raman spectroscopy. Of more interest to chemists, are such subjects as observation of phase change, high temperature and high pressure phenomena and comparison of the liquid state with the solid state.

Crystalline α-quartz has attracted the interest of several workers. Shapiro, O'Shea and Cummins[64] measured the Raman spectra of single-crystal α-quartz, in various orientations. The data were used to assign Raman-active frequencies. When the crystal was slowly heated to 559°C, the spectrum shifted over to that of β-quartz. The shift was found to be reversible. These workers explained some of the features of the α-phase spectrum by a "double-well" potential. Assell and Nichol,[65] using a diamond Raman cell, measured the spectra of α-quartz at 20 kbar and 40 kbar. They found that only those Raman bands shifted in frequency which shifted when the temperature was raised at normal pressure. They hypothesized that the shifts were due mainly to the change in interatomic distance. Melviger, Brasch and Lippincott studied the Raman spectra of liquid Br_2 and CS_2 and of single crystals formed at high pressure, in a diamond cell.[66] They found the 311 cm^{-1} stretching frequency of liquid Br_2 to split into two components, at 294 and 302 cm^{-1}, in the crystal. The position of the Raman bands of CS_2 at 646 and 655 cm^{-1}, were unaffected by crystallization. This was explained in terms of weak intermolecular interaction in CS_2.

Porto, Giordimaine and Damen[67] examined a single crystal of calcite, at various orientations. They were able to show that inconsistencies which appeared in spectra excited with mercury arc were caused by birefringence and convergence effects.

Assignments of frequencies based on single crystal studies have been given for $LiNO_3$[68], $CaWO_4$, $SrWO_4$, $CaMoO_4$ and $SrMoO_4$[69], sapphire[70], $SrTiO_3$,[71] and $LaBr_3$.[72]

Unfortunately, few studies in the extremely interesting area of mixed crystals seem to have appeared. One such study[73] was on mixed crystals of the type $(Ca_xSr_{x-1})F_2$ and $(Sr_xBa_{x-1})F_2$. It was found that the crystals had one Raman band each, whose position was a linear function of concentration. The bandwidth varied with concentration, reaching a maximum at $x = 0.5$. The bandwidth thus seemed to be a measure of disorder, for these crystals.

4.5   IONIC MELTS AND COMPLEX IONS IN SOLUTION

The study of the composition of ionic melts and of complex ions in solution has been greatly advanced by the application of Raman spectroscopy. Indeed, it is hard to see how any other method would yield such detailed information. A large body of literature, based on work with mercury arc excitation, exists.[R4] More recently studies, using modern laser Raman spectrometers, are giving much improved results. Besides the greater simplicity of the experiments, the laser Raman spectra provide improved resolution and accurate depolarization ratios and intensity data.

While work with solutions is perfectly straightforward, experiments on melts have their own peculiar difficulties. Most often, the melt is held in a quartz cuvette in a furnace. Care must be taken to prevent the spectrometer input optics from being damaged by heat. Melts which attack quartz must be held in a platinum or other inert metal boat, in an inert gas atmosphere. In this event, special optical arrangements must be used to get the exciting radiation into the sample and to collect the Raman radiation.

Three different cases may arise when an ionic solid is dissolved or melted.

a. The Raman spectrum is simply that of the constituent ions, say, that of NH_4^+ and NO_3^-.

b. The Raman spectrum is that of the constituent ions, but is somewhat perturbed. Quasi-lattice type bands may appear in the spectrum of the melt, degenerate vibrations may be split, etc.

c. Totally new species, such as CdI_4^{2-} in a solution of CdI_2 in water, may be formed.

Cases a and b are relatively easy to deal with. The geometrical structures of the ions are generally known in advance, so that the overall features of the spectrum can be predicted from group theory and standard chemical group frequencies. Any deviations from the predicted spectrum may be interpreted in terms of cation-anion pairing, residual local order, complexing with the solvent, etc. Case c is another matter. Here neither the number nor the nature of the complexes giving rise to the observed Raman spectrum is known a-priori. Before any progress can be made, the number of chemical species present must be sorted out, and the observed Raman bands assigned to the individual species. This is commonly done by measuring the integrated intensities of the Raman bands as a function of composition. For example, Janz and James[74] made up melts of $HgCl_2$ and KCl of different Hg^{2+}/Cl^- molar ratios and plotted the integrated intensities of the Raman bands as a function of composition. The intensities of bands originating from a given species changed in the same manner, as the composition changed. In the case cited, of the three strongest highly polarized Raman bands observed, one decreased in intensity, one went through a maximum and one increased in intensity, as the concentration of Cl^- was increased. This clearly implied a sequence

$$HgCl_2 \rightarrow HgCl_3^- \rightarrow HgCl_4^{2-}$$

The observed Raman bands were consistent with this interpretation.

Similar studies are routinely carried out in aqueous solution. In principle, the composition of simple complexes may be determined by the Job method.[R4] In this method, a series of solutions is prepared in which the sum of the molar concentrations of A and B is held constant, but A and B are varied. A and B are presumed to form a complex AB_n. If the intensity of a Raman band characteristic of the complex is plotted again $A/(A+B)$, the plot has a maximum at a value

$$\frac{A}{A+B} = f \tag{4.5.1}$$

It can be shown that

$$n = \frac{f}{1-f} \tag{4.5.2}$$

The method is more complicated and less reliable if more than one complex is present, or if n is large.

Equilibrium or dissociation constants can sometimes be estimated from the Raman spectra. It is necessary to have available some independent estimate of the quotient of activity coefficients, since only concentrations can be determined from the Raman spectrum. This approach is particularly useful if some of the species involved in the reaction can be prepared in pure form. Relative scattering coefficients for the Raman bands of the species involved can be measured on the pure species and used to determine concentrations in mixtures. The equilibrium between HSO_4^- and SO_4^{2-} in solution was studied by this method.[R4]

These studies are less useful in melts than in solution, since " intermolecular " forces are strong and often perturb the spectra. Melts of ionic substances often have low-lying Raman bands which can only be ascribed to quasi-lattice modes of the melt. These bands do not imply long-range order, but they would seem to imply some kind of long-range interactions such as those found in covalently bonded glasses. The selection rules for the known structures of ions such as OH^-, ClO_3^-, NO_3^-, etc., are reasonably well mirrored in the observed spectra. Violations, in the form of the splitting of degenerate modes, or the appearance of forbidden modes as weak Raman bands, do occur. A few applications of Raman spectroscopy to study the reaction kinetics of ionic species have been made.[R4] A promising recent development is in the use of Raman spectroscopy to make an *in situ* identification of $Zn(OH)_4^{2-}$ as the major electrolyte species in a battery electrolyte.[75]

The foregoing discussion outlined the use of Raman spectroscopy in the study of solution chemistry. It is worth mentioning that Raman spectroscopy is probably the quickest method available for identifying the complex anions or cations present in a solid sample or aqueous solution. While the method cannot detect, directly, elemental cations or anions, such as Cd^{2+} or I^-, these may be detected by causing them to complex, if, indeed, they form complexes. If, for example, Al^{3+} is suspected to be present, it may be detected by saturating the solution with KCl, which converts the Al^{3+} to $AlCl_4^{2-}$.

Table 4.1 lists the characteristic frequencies of some of the more common complex ions.

Table 4.1.   Raman Spectra of Some Complex Ions, cm^{-1}

$AuCl_4^-$	347	171	324	
$AuBr_4^-$	212	102	196	
ZnI_4^-	122	44	170	62
CdI_4^{2-}	117	36	145	44
HgI_4^{2-}	126	35		41
InI_4^-	139	42	185	58
$InBr_4^-$	197	55	239	79
$InCl_4^-$	321	89	337	112
$TlCl^{2+}$	327			
$TlCl_2^{1+}$	320			
$TlCl_3$	313			
$TiCl_4^-$	305	81		

A particularly interesting study is reported by Solomons, Clarke and Bockris.[76] These workers examined the Raman spectrum of molten cryolite at 1030°C, using a pulsed ruby laser for excitation and photographic recording of their spectra. They conclude that in molten cryolite, AlF_6^{3-} dissociates, to a considerable extent, into F^- and AlF_4^-. Hester and Scaife[77] studied melts of the composition $Zn(NO_3)_2 \cdot xH_2O$, with variable x. An analysis of the infrared absorption and laser Raman spectra led them to conclude that nitrate-zinc ion complexing increased with decreasing water content, until at very low water content no uncomplexed nitrate remained. Clarke and Hester[78] studied $InCl_3$-alkali metal chloride melts. They conclude that while $InCl_4^-$ is the only species present in LiCl melts, $InCl_5^{2-}$ and $InCl_6^{3-}$ are also present in KCl and CsCl melts. James and Leong[79] report Raman spectra of $LiNO_3$, $NaNO_3$ and $AgNO_3$ at 5°C and 50°C above the melting point. The spectra of the nitrate ions are strongly perturbed by ion interactions, the two degenerate vibrations both being split. To boot, all of the spectra exhibit strong "quasi-lattice" modes. The authors interpret the spectra on the basis of a cubic "quasi-lattice" existing in the melt.

Knoeck[80] reports on La(III)-nitrate complexes in aqueous solution. He interprets splitting of nitrate bands the infrared and Raman spectra as arising from ion-solvent interactions, new bands as being due to cation-anion complexes.

4.6 WATER, CHEMICAL ELEMENTS, AND SIMPLE, COVALENT INORGANIC COMPOUNDS

Raman spectroscopy has contributed and is continuing to contribute, a respectable fraction of our knowledge of the structure of liquid water. Walrafen has carried out extensive studies of the Raman spectrum of liquid water[81] and of HDO in H_2O.[82,83] Relative intensities of librational frequencies near 475 and 710 cm^{-1}, of the OH bending frequency near 1645 cm^{-1} and of the OH stretching frequencies near 3500 cm^{-1} were measured as a function of temperature. A figure of 2.5 ± 0.6 kcal per mole, for the heat of dissociation of a hydrogen bond, was calculated from these results.

An extremely interesting recent application of Raman spectroscopy is its use in determining the structure of "polywater."[84] The Raman spectrum differs from both the infrared absorption spectrum and the Raman spectrum of ordinary water. The authors conclude from the vibrational spectrum and other evidence, that the bonds of polywater are covalent bonds, much stronger than the usual hydrogen bonds.

Simple centrosymmetric molecules like N_2, Cl_2, etc., either have no infrared absorption spectrum, if they are diatomic or have part of the vibrational spectrum appear only in Raman scattering. Cahill and LeRoi[85] give the Raman spectra of solid CO_2, N_2O, and CO. They are able to show from the intermolecular lattice vibrations, that the latter three substances have crystal structures isomorphous with that of CO_2. They find the intra-molecular vibrations little affected by intermolecular interactions. The same authors[86] analyze the Raman spectra of the four phases of solid oxygen, and of solid Cl_2 and Br_2.[87] In the latter, they show that the intensities of Raman bands arising from each of the three expected isotopes are in accord with theoretical predictions. The Raman spectrum of polycrystalline ONF_3 is given by Abramowitz and Levin[88] and is used to make an assignment of frequencies to the normal modes. Durig, Antion and Pate[89] give the Raman spectra of solid PH_4Br and PD_4Br. These authors make an assignment of frequencies to lattice and internal vibrational modes, as well as an estimate of the barrier to free rotation of PH_4^+ and PD_4^+ ions.

Smith and Barrett[90] have published one of the few studies in inorganic chemistry which use the laser Raman spectra of a gas. They got excellent Raman spectra of gaseous carbon suboxide using 1 watt of exciting radiation at 5145 Å. It is interesting that they observe that the lowest lying vibrational band, an overtone at 127 cm^{-1}, falls within the rotational envelope (as, indeed, does the infrared active fundamental at 63 cm^{-1}). The results were used to clear up uncertainties in the assignment of vibrational frequencies.

4.7   ORGANIC   CHEMISTRY

The advent of laser Raman spectroscopy has had, so far, only a limited impact on the fields of organic or physical-organic chemistry. The reason for this is that there is already on hand a large body of literature from pre-laser-Raman days. The Raman spectra of most organic liquids or melts are rela-tively easy to record, so that useful results can be gotten with mercury-arc excitation. Materials which were photosensitive, fluorescent or only obtain-able as crystal powders were, of course, rarely examined. This is not to say that the bulk of the spectra reported in the literature will not have to be re-examined. Few spectra were recorded photoelectrically with mercury arc excitation, so that data on band intensities are practically nonexistent. Depolarization ratios are sometimes given, but these are seldom reliable. The intensity, depolarization ratio and halfwidth of a Raman band are important "tags" for characterizing the band, particularly in a mixture of compounds. Aside from this, there is some indication that they may be correlated with molecular structure, as is the frequency.

One of the most useful applications of Raman spectroscopy in organic chemistry is the study of conformational or rotational isomers. It is the general rule that only one isomer will be stable in the crystalline material and that, in the liquid, the various isomers will be in thermodynamic equilibrium. By temperature studies and by comparison of the spectra of solid and melt, bands associated with specific isomers may often be picked out. A parallel infrared absorption study is invaluable in this work. It goes without saying that other thermodynamic equilibria, such as those involving halogen-aromatic complexes, may be determined through Raman spectroscopy.

A reasonable amount of work is being done on the determination of matrix elements of the scattering tensor for Raman bands of organic single crystals. It is too early to say where this work may lead.

The spectra of both crystalline and matrix-isolated glyoxal have been studied by Vederame, Castellucci and Califano.[91] Their work included infrared spectra in the 70–4000 cm^{-1} range and laser Raman spectra at $-185°C$. They found a centrosymmetric s-trans structure in both crystalline and matrix-isolated material. They assigned D_{2h} symmetry to the crystal. Foglizzo and Novak[92] have made an interesting study of hydrogen bonding in pyridinum halides. They see $A^-...HB^+$ bonds in the 0–300 cm^{-1} region and the 1500–4000 cm^{-1} region. They conclude that the 0–300 cm^{-1} bands are due to stetchings, bendings and torsion of the $A^-...(HB^+)$ bond. Green and Harvey[93] on the basis of a study of $(CH_3)_2Se_2$ and $(CD_3)_2Se_2$ assign the C—Se stretching frequency at 584 cm^{-1}. A normal coordinate analysis was used to aid in the assignment of frequencies.

Infrared and Raman spectra of polycrystalline cyclopropane and cyclopropane d_6 were studied by Bates, Sands and Smith.[94] The data were used to assign frequencies and to select D_{2h} as the correct crystal symmetry from a number of possibilities suggested by X-ray diffraction work. Brown, Obremski, Alkins and Lippincott[95] made an analogous study on CH_2Cl_2 and CH_2Br_2. They assigned the normal frequencies and accounted for isotopic structure and lattice modes. It is interesting that in CH_2Br_2, most CH_2 modes are split, whereas in CH_2Cl_2 most CCl_2 modes are split. The authors explained this by supposing that in CH_2Br_2, mainly H...H repulsion predominates, whereas in CH_2Cl_2, dipole-dipole interactions predominate.

Stenman[96] reports laser Raman spectra of powdered benzil and anthroquinine, using the data to make partial assignments of frequencies. Creighton, Green and Harrison[97] give complete assignments of frequencies, based on infrared and Raman data, for some trimethyl sulfoxonium salts. The Raman spectra for this work were recorded on both crystal powders and aqueous solutions.

Hanson and Gee[98] report a detailed study of single crystals of napthalene, leading to an estimate of the elements of the scattering tensors. Unfortunately, the spectra were recorded photographically, so that the results are only semiquantitative. Suzuki and Ito[99] studied the laser Raman spectra of single crystals of p-dichlorobenzene and p-dibromobenzene. The data were used to assign frequencies and to estimate values of polarizability tensor elements, for various normal modes. Similar work on aryliodochlorides is reported by

Hikal, Wolf and Burg.[100] Infrared and Raman spectra of 2,5-dihydro-thiophene are given by Green and Harvey.[101] The data were used to assign frequencies to the normal modes of vibration. Nyquist[102] has published one of the few recent studies in which Raman spectra have been used to identify characteristic group frequencies. In his work in a group of organophosphorous compounds, he finds that in $(CH_3O)_2(P{=}O)H$, there is a strong $P{=}O$ stretching Raman band at 1265 cm^{-1}. Freeman and Mayo[103] report, in the first of what is to be a series of papers, on the Raman spectra of some acyclic terpenes. They report band positions, relative intensities and de-polarization ratios for the more important bands. These quantities are correlated with molecular structure. Saunders, Lucier and Willis[104] have made an extensive study of the 50–800 cm^{-1} region of the Raman spectra of ethylbenzene, n-propylbenzene and terminal halogen-substituted analogs. In addition to assignments of frequencies, they assign some bands to specific conformational isomers. Ginn, Hauqe and Wood[105] report infrared and Raman spectra of the complexes pyridine-bromine and pyridine-bromine chloride. They find that the symmetry of the un-ionized complex is the same as that of pyridine. The spectra are shown to arise from $PyBr_2$, $(Py_2Br)^+$ and Br_3^- or $BrCl_2^-$. They hypothesize an equilibrium $2PyBr_2 \rightleftharpoons (Py_2Br)^+ + Br_3^-$ with the un-ionized complex being favored. Perettie and Waite[106] have used the infrared and Raman spectra of difluoromalononitrile to assign frequencies and to calculate thermodynamic functions. Durig and Willis[107] determined the structure of silacyclopentane from the vibrational spectra. They concluded that the only element of symmetry possessed by the molecule is a twofold rotation axis.

The study of metalorganic compounds is of continuing interest. Durig, Sink, and Turner[108] report spectra of a series of triphenyl germanes. Their assignments of frequencies are aided by low-temperature Raman spectra. Another example is a study of some dimethyl and trimethyl lead halides.[109] The data are used to show that these compounds are monomers in benzene solution, but may be halogen-bridged polymers in the solid state. Durig, Lau, Turner and Bragen[110] and Durig, Craven and Bragen[111] report Raman spectra of a series of trimethyl and tri (deuteromethyl) carbon, silicon and germanium halides. Lattice modes were observed in infrared and Raman spectra of the polycrystalline materials. The Raman-active lattice modes broadened and shifted to higher frequency, with increasing temperature.

GENERAL REFERENCES

R1. H. Eyring, J. Walter, and G. E. Kimball, *Quantum Chemistry* (Wiley, New York, 1944).

R2. J. Brandmüller and H. Moser, *Einführung in die Raman Spektroskopie* (Steinkopf Verlag, Darmstadt, 1962).

R3. E. B. Wilson, Jr., J. C. Decius, and P. C. Cross, *Molecular Vibrations* (McGraw-Hill, New York, 1955).

R4. H. A. Szymanski, Ed., *Raman Spectroscopy* (Plenum Press, New York, 1967).

R5. G. Herzberg, *Molecular Spectra and Molecular Structure*, Vol. II, *Infrared and Raman Spectra of Polyatomic Molecules* (Van Nostrand, New York, 1945).

R6. H. Margenau and G. M. Murphy, The Mathematics of Physics and Chemistry, 2nd ed. (Van Nostrand, New York, 1956).

R7. G. F. Koster, *Space Groups and Their Representations* (Academic Press, New York, 1957).

R8. G. Herzberg, *Molecular Spectra and Molecular Structure*, Vol. I, *Spectra of Diatomic Molecules* (Van Nostrand, New York, 1950).

R9. W. J. Youden, *Statistical Methods for Chemists* (Wiley, New York, 1951).

R10. L. J. Bellamy, *Infrared Spectra of Complex Molecules* (Wiley, New York, 1958).

R11. N. B. Colthup, L. H. Daly, and S. E. Wiberley, *Introduction to Infrared and Raman Spectroscopy* (Academic Press, New York, 1964).

R12. F. F. Bentley, L. D. Smithson, and A. L. Rozek, *Infrared Spectra and Characteristic Frequencies* 700–300 cm^{-1} (Wiley, New York, 1968).

R13. K. W. F. Kohlrausch, *Ramanspektren* (Becker and Erler, Leipzig, 1943; reprinted by Edward Brothers, Ann Arbor, Mich., 1945).

R14. H. A. Szymanski, *Correlation of Infrared and Raman Spectra of Organic Compounds* (Hertillon Press, Cambridge Springs, Pa., 1969).

REFERENCES

1. M. C. Tobin, *Developments in Applied Spectroscopy*, Vol. I (Plenum Press, New York, 1962).
2. N. Ham and A. Walsh, *Spectrochimica Acta* **12**, 88 (1958).
3. S. P. S. Porto, *J. Opt. Soc. Am.* **56**, 1985 (1966).
4. M. C. Tobin, *J. Chem. Phys.* **23**, 891 (1955).
5. J. G. Skinner and W. G. Nilsen, *J. Opt. Soc. Am.* **58**, 113 (1968).
6. H. Winston and R. S. Halford, *J. Chem. Phys.* **16**, 1063 (1948).
7. S. P. S. Porto, P. A. Fleury, and T. C. Damen, *Phys. Rev.* **154**, 522 (1967); R. C. C. Leite, J. F. Scott, and T. C. Damen, *Phys. Rev. Letters* **22**, 780, 782 (1969). The theory is discussed by R. Loudon, *Advan. Phys.* **13**, 423 (1964).
8. N. Bloembergen, *Nonlinear Optics* (W. A. Benjamin, New York, 1965).
9. R. W. Terhune, P. D. Maker, and C. M. Savage, *Phys. Rev. Letters* **14**, 681 (1965).
10. S. J. Cyvin, J. E. Rauch, and J. C. Decius, *J. Chem. Phys.* **43**, 4083 (1965).
11. A. Maréchal, *Traité d'Optique Instrumentale*, Vol. I, *Imagerie Géométriqué. Aberrations* (Editions de la Revue d'Optique Theorique et Instrumentale, Paris, 1952) p. 14.
12. J. J. Barrett and N. I. Adams III, *J. Opt. Soc. Am.* **58**, 311 (1968).
13. M. Born and E. Wolf, *Principles of Optics* 3rd ed. (Pergamon Press, New York, 1965), p. 441.
14. M. C. Tobin, *J. Opt. Soc. Am.* **58**, 1057 (1968).
15. M. C. Tobin, *J. Chem. Phys.* **21**, 1110 (1953).
16. M. C. Tobin, *Appl. Opt.* **9**, 502 (1970).
17. J. K. Nakamura and S. E. Schwarz, *Appl. Opt.* **7**, 1073 (1968).
18. R. R. Alfano and N. Ockman, *J. Opt. Soc. Am.* **58**, 90 (1968).
19. Y. H. Pao, R. N. Zitter, and J. E. Griffiths, *J. Opt. Soc. Am.* **56**, 1133 (1966).
20. J. J. Barrett and A. Weber, *J. Opt. Soc. Am.* **60**, 70 (1970).
21. *Kodak Plates and Films* (Eastman Kodak Co., Rochester, N. Y.).
22. C. B. Neblette, *Photography—Its Materials and Processes*, 6th ed. (Van Nostrand, New York, 1962).
23. M. Bridoux, *Rev. Opt.* **46**, 389 (1967).
24. M. Delhaye, *Appl. Opt.* **7**, 389 (1967).
25. M. C. Tobin, *J. Opt. Soc. Am.* **49**, 850 (1959).
26. J. J. Barrett and M. C. Tobin, *J. Opt. Soc. Am.* **56**, 129 (1966).
27. E. E. Wahlstrom, *Optical Crystallography*, 4th ed. (Wiley, New York, 1969).
28. R. Loudon, *Advan. Phys.* **13**, 423 (1964).
29. L. N. Ovander, *Opt. Spectr.* **9**, 302 (1960).
30. T. C. Damen, S. P. S. Porto, and B. Tell, *Phys. Rev.* **142**, 570 (1966).
31. A. Van Valkenburg, *Appl. Opt.* **9**, 1 (1970).
32. P. J. Hendra and E. J. Loader, *Nature* **217**, 637 (1968).
33. E. R. Lippincott and M. C. Tobin, *J. Am. Chem. Soc.* **75**, 4141 (1953).
34. E. R. Lippincott, P. Mercier, and M. C. Tobin, *J. Phys. Chem.* **57**, 939 (1953).
35. M. C. Tobin, *J. Phys. Chem.* **61**, 1392 (1957).
36. L. C. Pauling, *Nature of the Chemical Bond*, 3rd ed. (Cornell University Press, Ithaca, N. Y., 1960).
37. N. B. Hannay and C. P. Smythe, *J. Am. Chem. Soc.* **68**, 171 (1946).
38. W. Gordy, *J. Chem. Phys.* **14**, 305 (1946).

39. M. C. Tobin, *J. Phys. Chem.* **64**, 216 (1960).
40. R. C. Lord and G. Thomas, *Spectrochim. Acta* **23A**, 2551 (1967).
41. M. C. Tobin, *J. Mol. Spectr.* **5**, 65 (1960).
42. S. W. Cornell and J. L. Koenig, *J. Appl. Phys.* **39**, 4883 (1968).
43. W. L. Peticolas, B. Fanconi, B. Tomlinson, L. A. Nafie, and W. Small, *Ann. N. Y. Acad. Sci.* **168**, 564 (1970).
44. B. Franconi, B. Tomlinson, L. A. Nafie, W. Small, and W. L. Peticolas, *J. Chem. Phys.* **51**, 3993 (1969).
45. P. J. Flory, *Principles of Polymer Chemistry* (Cornell University Press, Ithaca, N. Y., 1953).
46. M. C. Tobin, *J. Chem. Phys.* **50**, 4551 (1970).
47. D. S. Cain and A. B. Harvey, *Selected Commercial Polymers* **1968** [AD-681 718 USGRDR 69(7) 68 (1969)].
48. C. A. Frenzel, E. B. Bradley, and M. S. Mathur, *J. Chem. Phys.* **49**, 3789 (1968).
49. R. G. Snyder, *Mol. Spect.* **31**, 464 (1969).
50. R. F. Schaufele, *J. Opt. Soc. Am.* **57**, 105 (1967).
51. G. Zerbi and P. Hendra, *J. Mol. Spectr.* **30**, 159 (1969).
52. F. J. Boerio and J. L. Koenig, *J. Chem. Phys.* **50**, 2823 (1969).
53. H. Sugeta, T. Miyazawa, and T. Kajiura, *J. Polymer Sci., Part B—Polymer Letters* **7**, 251 (1969).
54. M. C. Tobin, *Science* **161**, 68 (1968).
55. M. C. Tobin, *Spectrochim. Acta* **25A**, 1855 (1969).
56. B. L. Tomlinson and W. L. Peticolas, *J. Chem. Phys.* **52**, 2154 (1970).
57. I. Simon, in *Modern Aspects of the Vitreous State*, J. D. MacKenzie, Ed. (Butterworths, Washington, D. C., 1960) p. 139.
58. W. P. Griffiths, *Nature* **224**, 264 (1969).
59. I. R. Beattie and T. R. Gilson, *Proc. Roy. Soc. (London)* **A307**, 407 (1968).
60. M. C. Tobin and T. Baak, *J. Opt. Soc. Am.* **58**, 1459 (1968).
61. M. C. Tobin and T. Baak, *J. Opt. Soc. Am.* **60**, 368 (1970).
62. R. J. Bell, N. F. Bird, and P. Dean, *J. Phys. C (Proc. Phys. Soc. London)* **1**, 299 (1968).
63. M. Hass, *Solid State Communications* **7**, 1069 (1969).
64. S. M. Shapiro, D. C. O'Shea, and H. A. Cummins, *Phys. Rev. Letters* **19**, 361 (1967).
65. J. F. Assell and M. Nicol, *J. Chem. Phys.* **49**, 5393 (1968).
66. A. J. Melveger, J. W. Brasch, and E. R. Lippincott, *Appl. Opt.* **9**, 11 (1970).
67. S. P. S. Porto, J. A. Giordmaine, and T. C. Damen, *Phys. Rev.* **147**, 608 (1966).
68. R. E. Miller, R. R. Getty, K. L. Treuil, and G. E. Leroi, *J. Chem. Phys.* **51**, 1385 (1969).
69. S. P. S. Porto and J. F. Scott, *Phys. Rev.* **157**, 716 (1967).
70. S. P. S. Porto and R. S. Krishnan, *J. Chem. Phys.* **47**, 1009 (1967).
71. W. G. Nilsen and J. G. Skinner, *J. Chem. Phys.* **48**, 2240 (1968).
72. C. K. Asawa, *Phys. Rev.* **173**, 869 (1968).
73. R. K. Chang, B. Lacina, and P. S. Pershan, *Phys. Rev. Letters* **17**, 755 (1966).
74. G. J. Janz and D. W. James, *J. Chem. Phys.* **38**, 905 (1963).
75. Private communication J. F. Jackovitz, Westinghouse Electric Corporation.
76. C. Solomons, J. H. R. Clark, and J. O'M. Bockris, *J. Chem. Phys.* **49**, 445 (1968).
77. R. E. Hester and C. W. J. Scaife, *J. Chem. Phys.* **47**, 5253 (1967).
78. J. H. R. Clarke and R. E. Hester, *J. Chem. Phys.* **50**, 3106 (1969).

79. D. W. James and W. H. Leong, *J. Chem. Phys.* **51**, 640 (1969).

80. J. Knoeck, *Analytical Chemistry* **41**, 2069 (1970).

81. G. E. Walrafen, *J. Chem. Phys.* **47**, 114 (1957).

82. G. E. Walrafen, *J. Chem. Phys.* **48**, 244 (1968).

83. H. S. Frank, *Science* **169**, 635 (1970).

84. E. R. Lippincott, R. R. Stromberg, W. H. Grant, and G. L. Cessac, *Science* **164**, 1482 (1969).

85. J. E. Cahill and G. E. Leroi, *J. Chem. Phys.* **51**, 1324 (1969).

86. J. E. Cahill and G. E. Leroi, *J. Chem. Phys.* **51**, 97 (1969).

87. J. E. Cahill and G. E. Leroi, *J. Chem. Phys.* **51**, 4514 (1969).

88. S. Abromowitz and L. W. Levin, *J. Chem. Phys.* **51**, 463 (1969).

89. J. R. Durig, D. J. Antion, and C. B. Pate, *J. Chem. Phys.* **51**, 4449 (1969).

90. W. M. Smith and J. J. Barrett, *J. Chem. Phys.* **51**, 1475 (1969).

91. F. D. Vederame, E. Catellucci, and S. Califano, *J. Chem. Phys.* **52**, 719 (1970).

92. R. Foglizzo and A. Novak, *J. Chem. Phys.* **50**, 5366 (1969).

93. W. H. Green and A. B. Harvey, *J. Chem. Phys.* **49**, 3586 (1968).

94. J. B. Bates, D. E. Sands, and W. H. Smith, *J. Chem. Phys.* **51**, 105 (1969).

95. C. W. Brown, R. J. Obremski, J. R. Alkins, and E. R. Lippincott, *J. Chem. Phys.* **51**, 1376 (1969).

96. F. Stenman, *J. Chem. Phys.* **51**, 3141, 3413 (1969).

97. J. A. Creighton, J. H. S. Green, and D. J. Harrison, *Spectrochim. Acta* **25A**, 1017 (1969).

98. D. M. Hanson and A. R. Gee, *J. Chem. Phys.* **51**, 5202 (1969).

99. M. Suzuki and M. Ito, *Spectrochim. Acta* **25A**, 1017 (1969).

100. A. Hikal, W. Wolf, and A. Burg, *Spectr. Letters* **2**, 13 (1969).

101. W. H. Green and A. B. Harvey, *Spectrochim. Acta* **25A**, 723 (1969).

102. R. A. Nyquist, *Spectr. Letters* **2**, 47 (1969).

103. S. K. Freeman and D. W. Mayo, *Appl. Spectr.* **23**, 610 (1969).

104. J. E. Saunders, J. J. Lucier, and J. N. Willis, *Spectrochim. Acta* **24A**, 2023 (1968).

105. S. G. W. Ginn, J. Hauque, and J. L. Wood, *Spectrochim. Acta* **24A**, 1531 (1968).

106. D. J. Perettie and S. C. Wait, Jr., *J. Mol. Spectr.* **32**, 222 (1969).

107. J. R. Durig and J. N. Willis, Jr., *J. Mol. Spectr.* **32**, 320 (1969).

108. J. R. Durig, C. W. Sink, and J. B. Turner, *Spectrochim. Acta* **25A**, 629 (1969).

109. R. J. H. Clark, A. J. Davies, and R. J. Puddenphatt, *J. Am. Chem. Soc.* **90**, 6923 (1968).

110. J. R. Durig, K. K. Lau, J. B. Turner, and J. Bragen, *J. Mol. Spectr.* **32**, 419 (1969).

111. J. R. Durig, S. M. Craven, and J. Bragen, *J. Chem. Phys.* **51**, 5663 (1969).

BIBLIOGRAPHY OF PUBLICATIONS ON RAMAN
SPECTROSCOPY, 1968–1969

Source. Spectra-Physics, Inc., 1250 West Middlefield Road, Mountain View, Calif. 94040. This is the first of a series of bibliographies which will be published periodically by Spectra-Physics and which may be obtained by writing to the company.

GENERAL PAPERS

R. O. Kagel, "Routine analytical applications of laser Raman spectroscopy," *Instrum News*, **19**, 7–8 (1968).

P. J. Hendra, "Laser-Raman spectrscopy applied to some chemical problems," *Mol. Spectrosc. Proc. Conf.*, 4th (1968), 285–97.

J. L. Koenig, "Laser Raman spectroscopy. Rebirth of a technique," *Res./Develop.* (Chicago), **20**, 18–20 (1969).

H. A. Szymanski, "The Role of Raman Spectra in Structural Analysis," *Progr. Infrared Spectrosc.*, **3**, 153–9 (1967). Review article.

J. C. Evans, "Advances in Raman spectroscopy," *Advan. Anal. Chem. Instrum.*, **7**, 41–66 (1968).

P. J. Hendra and P. M. Stratton, "Laser-Raman spectroscopy," *Chem. Rev.*, **69**, 325–44 (1969).

S. P. S. Porto, "Laser Raman spectroscopy," *Ind. Res.* **11**, 66–8 (1969).

I. R. Beattie and T. R. Gilson, "Single crystal laser Raman spectroscopy," *Proc. Roy. Soc.*, A, **307**, 407–29 (1968).

F. W. Karasek, "Table-top Raman spectroscopy," *Res./Develop. (Chicago)*, **20**, 32–3 (1969).

D. B. Powell, "Infrared and Raman Spectroscopy," *Annu. Rep. Progr. Chem.*, **63**, 112–28 (1966). Review article.

J. H. Parker, Jr., "Argon ion laser source for Raman scattering studies," *U. S. Govt. Res. Develop. Rep.*, **69**, 67 (1969).

P. B. Miller and J. D. Axe, "Internal strain and Raman active vibrations in solids." *Phys. Rev.*, **163**, 924–6 (1967).

P. J. Hendra, "Infrared and Raman spectroscopy," *Annu. Rep. Progr. Chem.*, **64**, 189–203 (1967).

B. F. Rud'ko, N. A. Krinitsyna, and V. G. Moskalinko, "Use of Raman spectral analysis to determine the concentration of the γ-isomer of hexachlorocyclohexane," *Zh. Prikl., Spectrosk.*, **6**, 783–8 (1967) (Russ.) *C.A.* **68**: 100391a.

F. Carre and R. Corriu, "Mechanism of the C-acylation of aromatic and ethylenic compounds. V. Structure of acyl chlorides and chlorides and sulfuric acid solutions. Reaction with cyclohexene," *Bull. Soc. Chim. Fr.*, 2898–904 (1967).

J. C. Laufer, J. E. Cahill, and G. E. Leroi, "Lattice vibrations in some simple molecular crystals," *Nat. Bur. Stand. (U. S.), Spec. Publ.*, **301**, 327–9 (1967).

I. J. Bryant, "Cryostat for the measurement of low temperature Raman spectra of crystals," *Spectrochim. Acta.*, **24A**, 9–14 (1968).

C. H. Ting, "Polarized Raman spectra, selection rules," *Spectrochim. Acta.*, **24A**, 1177–89 (1968).

J. A. Konigstein, "Theory of Raman scattering for overtone and combination bands in the vibrational Raman effect," *J. Mol. Spectros.*, **28**, 209–15 (1968).

G. Fini, P. Mirone and P. Patella, "Solvent effects on Raman band intensities," *J. Mol. Spectros.*, **28**, 144–60 (1968).

SMALL SAMPLE TECHNIQUES

S. K. Freeman and D. O. Landon, "Small-sample handling in laser Raman spectrometry," *Anal. Chem.*, **41**, 398–400 (1969).

T. Kajiura, "Laser-excited Raman spectra of microsamples," *Bunseki Kagaku*, **17**, 632–4 (1968) (Japan).

SPECIAL STUDIES

D. W. Feldman, J. H. Parker, Jr., and M. Ashkin, "Raman scattering by optical modes of metals," *Phys. Rev. Lett.*, **21**, 607–8 (1968).

R. C. C. Leite and J. F. Scott, "Resonant surface Raman scattering in direct gap semiconductors," *Phys. Rev. Lett.*, **22**, 130–2 (1969).

S. Pinchas, "Effect of hydrogen bonding on the stretching frequencies of oxygen-18-labelled ions," *J. Chem. Phys.*, **51**, 2284–5 (1969).

I. W. Shepard, A. R. Evans, and D. B. Fitchen, "Impurity-induced Raman spectra of nitrite-doped potassium iodide," *Phys. Lett. A*, **27**, 171–2 (1968).

A. Oseroff and P. S. Pershan, "Raman scattering from localized magnetic excitations in Ni^{++}- and Fe^{++}-doped manganous fluoride," *Phys. Rev. Lett.*, **21**, 1593–6 (1968).

W. Holzer, W. F. Murphy, and H. J. Bernstein, "Raman spectra of negative molecular ions doped in alkali halide crystals," *J. Mol. Spectrosc.*, **32**, 13–23 (1969).

G. Fini, P. Mirone, and P. Patella, "Solvent effects on Raman band intensities," *J. Mol. Spectrosc.*, **28**, 144–60 (1968).

RAMAN SPECTRA OF ADSORBED SPECIES

E. V. Pershina, and Sh. Sh. Raskin, "Certain features of Raman spectra of absorbed species." *Opt. Spektrosk. Akad. Nauk, SSSR, Otd. Fiz-Mat. Nauk, Sb. Stalei*, **3**, 328–37 (1967) (Russ). *CA* **68**, 109980z.

P. J. Hendra and E. J. Loader, "Laser Raman spectra of sorbed species. Physical adsorption on silica gel," *Nature*, **216**, 789–90 (1967).

P. J. Hendra and E. J. Loader, "Laser Raman Spectra of Sorbed Species," *Nature*, **217**, 637–8 (1968).

RAMAN SPECTRA OF GASES

M. Berjot, L. Bernard, and R. Dupeyrat, "Raman Effect in Gases," *Ann. Phys. (Paris)*, **2**, 293–303 (1967).

D. L. Renschler, J. L. Hunt, T. K. McCubbin, Jr., and S. R. Polo, "Rotational Raman spectrum of nitrous oxide," *Mater. Res. Bull.*, **4**, 551–3 (1969).

D. L. Renschler, J. L. Hunt, T. K. McCubbin, and S. R. Polo, "Rotational Raman spectrum of nitric oxide," *J. Mol. Spectros.*, **32**, 347–50 (1969).

W. H. Smith and J. J. Barrett, "Gas phase Raman spectra of carbon suboxide," *J. Chem. Phys.*, **51**, 1475–9 (1969).

H. H. Classen, H. Selig, and J. Shamir, "Raman apparatus using laser excitation and polarization measurements. Rotational spectrum of fluorine," *Appl. Spectrosc.* **23**, 8–12 (1969).

M. Brith, A. Ron, and O. Schnepp, "Raman Spectrum of α-N_2," *J. Chem. Phys.*, **51**, 1318–23 (1969).

I. R. Beattie and G. A. Ozin, "Gas-Phase Raman spectroscopy of trigonal bipyramidal pentachlorides and pentabromides," *J. Chem. Soc. A.* 1691–3 (1969).

J. J. Barrett and N. I. Adams, III, "Laser-excited rotation-vibration Raman scattering in ultrasmall gas samples," *J. Opt. Soc. Amer.*, **58**, 311–19 (1968).

I. I. Kondilenko, G. A. Vorobeva, and I. F. Klassen, "Fermi resonance in Raman spectra in liquid and gas states of a substance," *Opt. Spektrosk*, **27**, 420–5 (1969).

D. E. Shaw and H. L. Welsh, "Raman spectrum of ethane-1, 1, 1, d_3," *Can. J. Phys.*, **45**, 3823–35 (1967).

RAMAN SPECTRA OF CONDENSED GASES

J. E. Cahill and G. E. Leroi, "Raman studies of molecular motion in condensed oxygen," *J. Chem. Phys.*, **51**, 97–104 (1969).

J. E. Cahill and G. E. Leroi, "Raman spectra of solid carbon dioxide, nitrous oxide, nitrogen, and carbon monoxide," *J. Chem. Phys.*, **51**, 1324–32 (1969).

J. E. Cahill, K. L. Tremil, R. E. Miller, and G. E. Leroi, "Raman spectra of polycrystalline carbon dioxide and nitrous oxide," *J. Chem. Phys.*, **47**, 3678–9 (1967).

GEOLOGICAL SAMPLES

W. P. Griffith, "Raman studies on rock-forming minerals. I. Orthosilicates and cyclosilicates," *J. Chem. Soc. A*, 1372–7 (1969).

R. Forneris, "Infrared and Raman spectra of realgar and orpiment," *Amer. Mineral.*, **54**, 1062–74 (1969).

SAMPLES IN AQUEOUS SOLUTION

M. P. Hanson and R. A. Plane, "Raman spectroscopic study of stepwise bromide and chloride complexes of indium (III) in aqueous solution," *Inorg. Chem.*, **8**, 746–50 (1969).

A. R. Davis and D. E. Irish, "Infrared and Raman spectral study of aqueous mercury (II) nitrate solutions," *Inorg. Chem.*, **7**, 1699–704 (1968).

R. P. J. Cooney and J. R. Hlall, "Raman spectra of mercury (II) nitrate in aqueous solution and as the crystalline hydrate," *Aust. J. Chem.*, **22**, 337–45 (1969).

A. R. Davis, "Vibrational spectroscopic study of aqueous cadmium nitrate solutions," *Inorg. Chem.*, **7**, 2565–9 (1968).

T. M. Loehr and R. A. Plane, "Raman spectra and structures of arsenious acid and arsenites in aqueous solution," *Inorg. Chem.*, **7**, 1708–14 (1968).

D. E. Irish and A. R. Davis, "Interactions in Aqueous Alkali Metal Nitrate Solutions," *Can. J. Chem.*, **46**, 943–51 (1968).

A. Commeyras and G. A. Olah, "Chemistry in super acids. II. Nuclear magnetic resonance and laser Raman spectroscopic study of the antimony pentafluoride-fluorosulfuric acid (sulfur dioxide) solvent system (magic acid). The effect of added halides, water, alcohols, and carboxylic acids. Study of the hydronium ion," *J. Amer. Chem. Soc.*, **91**, 2929–42 (1969).

L. A. Blatz and P. Waldstein, "Low frequency Raman spectra of aqueous solutions of formates and acetates," *J. Phys. Chem.*, **72**, 2614–18 (1968).

S. Pinchas and D. Sadeh, "Fundamental vibration frequencies of the main isotopic $(PO_4)^{3-}$ ions in aqueous solutions," *J. Inorg. Nucl. Chem.*, **30**, 1785–9 (1968).

A. W. Herlinger and T. V. Long, "Investigation of the structure of the di-sulfate ion in aqueous solution using Raman and infrared spectroscopies," *Inorg. Chem.*, **8**, 2661–5 (1969).

T. M. Loehr and R. A. Plane, "Raman spectra of arsenic trichloride in water and alcohols and the spectrum of arsenic tribromide," *Inorg. Chem.*, **8**, 73–8 (1969).

METAL-METAL BONDS

D. M. Adams, J. N. Crosby, and R. D. W. Kenmitt, "Vibrational spectra of some metal-metal bonded compounds," *J. Chem. Soc. A*, 3056–8 (1968).

D. M. Adams, J. B. Cornell, J. L. Dawes, and R. D. W. Kemmitt, "Metal-metal stretching frequencies in trinuclear systems," *Inorg. Nucl. Chem. Lett.*, **3**, 437–9 (1967).

H. Stammreich and T. Teixeira Sans, "Mercury-mercury stretching frequencies and bond lengths in mercurous compounds," *J. Mol. Struct.*, **1**, 55–60 (1967).

C. O. Quicksall and T. G. Spiro, "Metal-metal frequencies and force constants of tetrairidiumdodecacarbonyl," *Inorg. Chem.*, **8**, 2011–13 (1969).

C. O. Quicksall and T. G. Spiro, "Raman frequencies and metal-metal force constants for di-metal deca-carbonyl species," *Inorg. Chem.*, **8**, 2363–7 (1969).

POLYMERS

R. F. Schaufele, "Advances in vibrational Raman scattering spectroscopy of polymers," *Trans. N. Y. Acad. Sci.*, **30**, 69–80 (1967).

R. F. Schaufele and T. Shimanouchi, "Longitudinal acoustical vibrations of finite polymethylene chains," *J. Chem. Phys.*, **47**, 3605–10 (1967).

D. S. Cain and A. B. Harvey, "Raman spectroscopy of polymeric materials. I. Selected commercial polymers," *U. S. Govt. Res. Develop. Rep.*, **69**, 68 (1969).

S. W. Cornell and J. L. Koenig, "Laser-excited Raman scattering in polystyrene," *J. Appl. Phys.*, **39**, 4883—90 (1968).

J. L. Koenig and F. J. Boerio, "Raman scattering in poly (vinyl fluoride)," *Makromol. Chem.*, **125**, 302–5 (1969).

J. L. Koenig and D. Druesedow, "Raman spectra of extended-chain syndiotactic poly (vinyl chloride)," *J. Polym. Sci.*, *Part A-2*, **7**, 1075–84 (1969).

J. L. Koenig and F. J. Boerio, "Raman scattering and band assignments in poly-(tetrafluoroethylene)," *J. Chem. Phys.*, **50**, 2823–8 (1969).

A. C. Angood and J. L. Koenig, "Laser-excited Raman spectra of poly(ethylene sulfide)," *J. Macromol. Sci. Phys.*, **3**, 323–9 (1969).

P. J. Hendra and H. A. Willis, "Laser-excited Raman spectra of stretched polyethylene fibers," *Chem. Commun.*, **1968**, 225–7.

R. G. Snyder, "Raman spectrum of polyethylene and the assignment of the B_{2g} wag fundamental," *J. Mol. Spectrosc.*, **31**, 464–5 (1969).

C. A. Frenzel, E. B. Bradley, and M. S. Mathur, "Raman spectrum of a crystalline polyethylene," *J. Chem. Phys.*, **49**, 3789–91 (1968).

G. Zerbi and P. J. Hendra, "Laser-excited Raman spectra of polymers: syndiotactic polypropylene," *J. Mol. Spectrosc.*, **30**, 159–62 (1969).

G. Zerbi and P. J. Hendra, "Laser-excited Raman spectra of polymers: hexagonal and orthorhombic poly(oxymethylene)," *J. Mol. Spectros.*, **27**, 17–26 (1968).

H. Sugeta, T. Miyazawa, and T. Kajiura, "Laser Raman scattering of poly(oxymethylene)," *J. Polym. Sci.*, *Part B*, **7**, 251–3 (1969).

M. C. Tobin, "Laser Raman spectra of polymethacrylic acid," *J. Chem. Phys.*, **50**, 4551–4 (1969).

W. R. Feairheller, Jr., and J. E. Katon, "The vibrational spectra and structure of poly-methylacrylate and vinylacetate," *J. Mol. Struct.*, **1**, 239–48 (1968).

E. R. Lippincott, R. R. Stromberg, W. H. Grant, and G. L. Cessac, "Polywater," *Science*, **164**, 1482–7 (1969).

L. J. Bellamy, A. R. Osborn, E. R. Lippincott, and A. R. Bandy, "Studies of the molecular structure and spectra of anomalous water," *Chem. Ind.* (*London*), 686–8 (1969).

P. J. Hendra, "Laser Raman spectra of polymers," *Fortschs. Hochpolym. Forsch.*, **6**, 151–69 (1969) (Eng.).

L. R. Maksanova, T. C. Godovikova, S. S. Novikov, and V. A. Shlypochnikov, "Polymerization of acrylates and methacrylates of various alcohols," *Tr. Vost. Sib. Tekhnol. Inst.*, **1**, 109–18 (1966). (Russ.).

P. J. Hendra, "B_{2g} methylene wagging mode in polyethylene," *J. Mol. Spectros.*, **28**, 118–19 (1968).

M. Tasumi and G. Zerbi, "Vibrational analysis of random polymers," *J. Chem. Phys.*, **48**, 3813–20 (1968).

M. J. Hannon, F. J. Boerio, and J. L. Koenig, "Vibrational analysis of poly(tetra-fluoroethylene)," *J. Chem. Phys.*, **50**, 2829–36 (1969).

R. F. Schaufele, "Chain shortening in polymethylene liquids," *J. Chem. Phys.*, **49**, 4168–75 (1968).

J. L. Koenig and D. Druesedow, "Raman spectra of extended chain syndiotactic poly(vinylchloride)," *J. Polym. Sci., Part A-2*, **7**, 1075–84 (1969).

P. J. Hendra, J. R. Mackenzie, and P. Holliday, "Laser Raman spectrum of poly (vinylidine chloride)," *Spectrochim. Acta*, **25A**, 1349–54 (1969).

S. W. Cornell, and J. L. Koenig, "Raman spectra of polybutadiene rubbers," *Macromol.*, **2**, 540–5 (1969).

S. W. Cornell and J. L. Koenig, "Raman spectra of polyisoprene rubbers," *Macromol.*, **2**, 546–9 (1969).

J. L. Koenig and F. J. Boerro, "Raman scattering in nonplanar poly(vinylidene fluoride)," *J. Polym. Sci., Part A-2*, **7**, 1489–94 (1969).

BIOLOGICAL SAMPLES

B. M. Fanconi, B. Tomlinson, L. A. Nafie, W. Small, and W. L. Peticolas, "Polarized laser Raman studies of biological polymers," *J. Chem. Phys.*, **51**, 3993–4005 (1969).

R. C. Lord and G. J. Thomas, Jr., "Spectroscopic studies of molecular interaction in DNA constituents," *Develop. Appl. Spectrosc.*, **6**, 179–99 (1967).

K. Hirano, "Raman spectra of DNA in aqueous solution," *Bull. Chem. Soc. Jap.*, **41**, 731–2 (1968) (Eng.). *CA*, **68**: 101913*j*.

L. Rimai, T. Cole, J. L. Parsons, J. T. Hickmott, Jr., and E. B. Carew, "Raman spectra of water solutions of adenosine tri-, di-, and monophosphate and some related compounds," *Biophys. J.*, **9**, 320–9 (1969).

M. C. Tobin, "Raman spectra of crystalline lysozyme, pepsin, and α-chymotrypsin," *Science*, **161**, 68–9 (1968).

J. L. Koenig and P. L. Sutton, "Raman spectrum of the right handed α-helix of poly-1-alanine," *Biopolymers*, **8**, 167–71 (1969).

M. Smith, A. G. Walton, and J. L. Koenig, "Raman spectroscopy of poly-l-proline in aqueous solution," *Biopolymers*, **8**, 173–9 (1969).

M. Smith, A. G. Walton, and J. L. Koenig, "Raman spectrum of polyglycines," *Biopolymers*, **8**, 29–43 (1969).

QUARTZ, SILICATES, AND GLASSES

M. Hass, "Temperature dependence of the Raman spectrum of vitreous silica," *Solid State Commun.*, **7**, 1069–71 (1969).

M. Boffe, J. Creogaert, and V. Caron, "Demixing in the silicon dioxide-sodium oxide-calcium oxide system. Application of Raman spectroscopy to the study of silica and some vitreous silicates," *Silicates Ind.*, **34**, 137–43 (1969) (Fr.).

W. Wadia and L. S. Ballomal, "Interpretation of the vibrational spectra of fused silica," *Phys. Chem. Glasses*, **9**, 115–24 (1968).

J. F. Asell and M. Nicol, "Raman spectrum of α-quartz at high pressures," *J. Chem. Phys.* **49**, 5395–9 (1968).

M. C. Tobin and T. Baak, "Raman spectra of some low-expansion glasses," *J. Opt. Soc. Amer.*, **58**, 1459–61 (1968).

R. E. Hester and K. Krishnan, "Vibrational spectra of some glasses of alkali-metal nitrates with Group IIA and IIB metal nitrates and with zinc chloride and water," *J. Chem. Soc. A*, 1955–60 (1968).

TEMPERATURE AND PRESSURE EFFECTS

D. J. Antion and J. R. Durig, "Variable temperature Raman cell," *Appl. Spectrosc.*, **22**, 675–7 (1968).

G. Seillies, M. Ceccaldi, and J. P. Leicknam, "Raman cell permitting work under pressure with ordinary and deuterated liquid ammonia," *Method. Phys. Anal.*, **4**, 388–95 (1968) (Fr.).

R. Cavagnat, J. J. Martin, and G. Turrell, "Laser-Raman cell for pressurized liquids," *Appl. Spectrosc.*, **23**, 172–3 (1969).

J. W. Brasch, A. J. Melveger, and E. R. Lippincott, "Laser excited Raman spectra of samples under very high pressure," *Chem. Phys. Lett.*, **2**, 99–100 (1968).

O. Brafman, S. S. Mitra, R. K. Crawford, W. B. Daniels, C. Postmus, and J. R. Ferraro, "Pressure dependence of Raman spectra of solids. Phase transition in thallous iodide," *Solid State Commun.*, **7**, 449–52 (1969).

ORGANOMETALLIC COMPOUNDS

David M. Adam, *Metal Ligand and Related Vibrations. A Critical Survey of the Infrared and Raman Spectra of Metallic and Organometallic Compounds* (Arnold, London, 1967).

R. J. H. Clark and B. C. Crosse, "Far-infrared and Raman spectra of halogeno-metal pentacarbonyls," *J. Chem. Soc. A*, 224–8 (1969).

R. L. Smith, "Low frequency vibrational spectra of group IVa phenyl compounds," *Spectrochim. Acta*, **24A**, 695–706 (1968).

M. G. Miles, J. H. Patterson, C. W. Hobbs, M. J. Hopper, J. Overend, and R. S. Tobias, "Raman and infrared spectra of isosteric diammine and dimethyl complexes of heavy metals. Normal-coordinate analysis of $(X_3Y_2)_2Z$ ions and molecules," *Inorg. Chem.*, **7**, 1721–9 (1968).

K. Krishnan and R. A. Plane, "Raman spectra of ethylenediaminetraacetic acid and its metal complexes," *J. Amer. Chem. Soc.*, **90**, 3195–200 (1968).

. H. S. Green, W. Kynaston, and G. A. Rodley, "Vibrational spectra of ligands and complexes. III. Dimethylphenylarsine, methyldiphenylarsine, triphenylarsine, and o-phenylenebis (dimethylarsine)," *Spectrochim. Acta*, **24A**, 853–62 (1968).

T. Ogewa, "Vibrational assignments and normal vibrations of trimethylaluminum," *Spectrochim. Acta*, **24A**, 15–20 (1968).

J. Weidlein and V. Krieg, "Vibrational spectra of dimethyl and diethylaluminum fluoride," *J. Organometal. Chem.*, **11**, 9–16 (1968).

J. H. Carpenter, W. J. Jones, R. W. Jotham, and L. H. Long, "Laser-source Raman spectroscopy and the Raman spectra of the methyldiboranes," *Chem. Commun.*, **15**, 881–3 (1968).

R. A. Walton, "Complexes of alkyl and aryl cyanides. VI. The vibrational spectra of $MX_2 \cdot 2RCN$, where M = palladium or platinum, X = chloride or bromide, and R = methyl or phenyl and the assignment of the metal-nitrogen stretching frequency," *Can. J. Chem.*, **46**, 2347–52 (1968).

D. A. Brown, D. Cunningham, and W. K. Glass, "Chromium (III) alkoxides," *J. Chem. Soc.*, A, 1563–8 (1968).

H. J. Buttery, G. Keeling, S. F. A. Kettle, I. Paul, and P. J. Stamper, "Solid-state studies. II. Raman- and infrared-active carbonyl stretching vibrations of four methylbenzenetri-carbonyl-chromium complexes," *J. Chem. Soc.*, A, 2224–7 (1969).

H. J. Buttery, G. Keeling, S. F. A. Kettle, C. Paul, and P. J. Stamper, "I. Raman and infrared-active carbonyl stretching vibrations of π-benzenetricarbonylchromium," *J. Chem. Soc.*, A, **14**, 2079–80 (1969).

R. A. Nyquist, "Vibrational spectroscopic studies of organophosphorus compounds: $ClCH_2-PCl_2$, $XCH_2-P(=O)X_2$, and $ClCH_2-P(=S)-Cl_2$," *Appl. Spectrosc.*, **22** Pt(1), 425–9 (1968).

I. R. Beattie and G. A. Ozin, "Adducts of group III trihalides with trimethylamine and trimethylphosphine," *J. Chem. Soc.* A, 2737–7 (1968).

J. R. Durig, J. S. DiYorio, and W. D. Wertz, "Vibrational spectra and structure of organophosphorous compounds. V. Infrared and Raman spectra of triethylphosphine sulfide and triethylphosphine selenide," *J. Mol. Spectrosc.*, **28**, 444–53 (1968).

P. J. D. Park and P. J. Hendra, "Vibrational spectra of square phanar complexes of formula *trans*-MX_2Y_2, where M = platinum (II) or palladium (II), X = chlorine, bromine, or iodine, and Y = trimethylphosphine or trimethylarsine," *Spectrochim. Acta*, **25A**, 909–16 (1969).

D. A. Duddell, P. L. Goggin, R. J. Goodfellow, and M. G. Norton, "The relationship of skeletal stretch frequencies throughout the trimethylphosphineplatinum (II) halide substitutional series," *Chem. Commun.*, **15**, 879–81 (1968).

I. R. Beattie and K. M. S. Livingston, "Single-crystal Raman spectrum of *trans*-hydridochlorobis(triethylphosphine)platinum," *J. Chem. Soc.* A, 2201–3 (1969).

D. E. Clegg and J. R. Hall, "Vibrational spectrum of hydroxytrimethylplatinum (IV)," *J. Organometal. Chem.*, **17**, 175–8 (1969).

J. Hiraishi, "Vibrational spectra of several platinum-ethylene complexes: $K(PtCl_3(C_2H_4)) \cdot H_2O$(Zeise's salt), $K(PtCl_3(C_2D_4)) \cdot H_2O$, and $(PtCl_2(C_2H_4))_2$," *Spectrochim. Acta*, **25A**, 749–60 (1969).

H. Buerger and W. Kilian "Spectroscopic studies of tris(trimethylsilyl)silane and tris(trimethylsilyl)silane-*d*," *J. Organometal. Chem.*, **18**, 299–306 (1969) (Ger).

H. Buerger, U. Goetze, and W. Sawodny, "Vibrational spectra and force constants of silyl and trimethylsilyl compounds of Group VI elements," *Spectrochim. Acta*, **24A**, 2003–13 (1968) (Ger.).

H. Buerger and W. Sawodny, "Vibrational spectra and force constants of methylsilyl and chlorosilyl dialkylamines," *Spectrochim. Acta*, **23A**, 2827–39 (1967).

K. G. Allum, J. A. Creighton, J. H. S. Green, G. J. Minkoff, and L. J. S. Prince, "Vibrational spectra of some dialkyl and diaryl disulfides and of dibutyl diselenide," *Spectrochim. Acta*, **24A**, 927–41 (1968).

E. Benedetti and V. Bertini, "Infrared and Raman spectra of 1,2,5-selenadiozol, *Spectrochim. Acta*, **24A**, 57–63 (1968).

G. C. Hayward and P. J. Hendra, "Infrared and laser Raman spectra of some species of the type Me_2YX_2 when Y = sulfur, selenium or tellurium and X is a halogen atom," *J. Chem. Soc.*, A, 1760–4 (1969).

D. Hartley and M. J. Ware, "Vibrational spectra and assignment of bis(π-cyclo-pentadienyl)cobalt(III) cation," *J. Chem. Soc. A*, 138–42 (1969).

J. N. Willis, Jr., M. T. Ryan, F. L. Hedberg, and H. Rosenberg, "Infrared and Raman study of 1,1'-disubstituted ferrocene compounds," *Spectrochim. Acta*, **24A**, 1561–72 (1968).

J. Bodenheimer, E. Loewenthal, and W. Loew, "Raman spectra of ferrocene," *Chem. Phys. Lett.*, **3**, 715–16 (1969).

T. V. Long, Jr., and J. R. Hnege, "The laser-Raman spectrum of ferrocene," *Chem. Commun.*, 1239–41 (1968).

W. Steingross and W. Zeil, "Preparation and vibrational spectra of 1-trimethylsilyl-2-chloroacetylene, 1-trimethylgermanyl-2-chloroacetylene, 1-trimethylstannyl-2-chloroacetylene, 1-trimethylplumbyl-2-chloroacetylene," *J. Organometal. Chem.*, **6**, 109–16 (1966).

J. R. Durig and C. W. Sink, "The vibrational spectra and structure of organoger-manes. II. Pentadeuterophenyltrichlorogermane," *Spectrochim. Acta*, **24A**, 575–87 (1968).

J. R. Durig, B. M. Gibson, and C. W. Sink, "Vibrational Spectra and Structure of Organogermanes. III. $C_6H_5GeBr_3$ and $C_6D_5GeBr_3$", *J. Mol. Struct.*, **2**, 1–17 (1968).

W. S. Cradock, "Germyl chemistry. VII. Vibrational spectra of digermylether, methoxygermane, and methylthiogermane," *J. Chem. Soc. A*, 1426–31 (1968).

J. R. Durig, J. B. Turner, B. M. Gibson, and C. W. Sink, "Vibrational spectra and structure of organogermanes. VII. Low wavenumber vibrations of some di-phenylgermanes," *J. Mol. Struct.*, **4**, 79–89 (1969).

R. A. Walton, "Raman spectra of crystalline complexes of indium (III) chloride with 2,2'-bipyridine, 1, 10-phenanthroline, pyridine, and pyrazine," *J. Chem. Soc. A*, 61–5 (1969).

I. R. Beattie and G. A. Ozin, "Single-crystal Raman spectrum of bis (trimethyl-amine)trichloroindium (III)," *J. Chem. Soc. A*, 542–5 (1969).

R. J. H. Clark, A. G. Davis, and R. J. Puddephatt, "Vibrational spectra and struc-ture of organolead compounds. II. Tetraphenyllead, hexaphenyldilead, tri-phenyllead halides, and diphenyllead dihalides," *Inorg. Chem.*, **8**, 456–63 (1969).

R. J. H. Clark, A. G. Davies, R. J. Puddephatt, and W. McFarlane, "Infrared, Raman and hydrogen-1, carbon-13, and lead-207 nuclear magnetic resonance spectra of hexamethyldilead," *J. Amer. Chem. Soc.*, **91**, 1334–9 (1969).

J. H. R. Clarke and L. A. Woodward, "Raman spectrophotometric studies of the dissociations of methylmercuric methanesulfonate and methylmercuric sulfate," *Trans. Faraday Soc.*, **64**, 1041–51 (1968).

M. J. Brookes and N. Jonathan, "Structure of trinitromethane and its covalent mercury salt," *J. Chem. Soc. A*, 2266–9 (1968).

D. Seybold and K. Dehnicke, "Vibrational spectra of some dialkyl and aryl mercury compounds," *J. Organometal. Chem.*, **11**, 1–8 (1968).

Z. Meic and M. Kandic, "Force constant calculations of methyl mercuric bromide and methyl-d_3 mercuric bromide," *Trans. Faraday Soc.*, **64**, 1438–46 (1968).

A. Buckingham and R. P. H. Gasser, "Complex formation in dimethylsulfoxide. II. Mercuric iodide and potassium iodide," *J. Chem. Soc. A*, 1964–5, (1967).

E. Maslowsky and K. Nakamoto "Infrared, Raman and proton magnetic resonance spectra of dicyclopentadienylmercury and related compounds," *Inorg. Chem.*, **8**, 1108–131 (1969).

P. J. D. Park and P. J. Hendra, "Vibration spectrum of trimethylphosphine-d_9," *Spectrochim. Acta*, **24A**, 2081–7 (1968).

J. R. Durig and J. S. DiYorio, "Vibrational spectra and structure of organophosphorus compounds. VII. Infrared and Raman spectra of trimethyl thionophosphate," *J. Mol. Struct.*, **3**, 179–90 (1969).

R. A. Nyquist, "Vibrational spectroscopic study of $(Me_2PO_2)^-$ and $(R_2PO_2)^-$," *J. Mol. Struct.*, **2**, 111–22 (1968).

R. A. Nyquist and W. W. Muelder, "Vibrational spectroscopic study of *o*-(2-propynyl)-phosphorochloridothioate," *J. Mol. Struct.*, **2**, 465–73 (1968).

M. Webster and M. J. Deveny, "Reaction of 1, 10-phenanthroline with some phosphorus(V), arsenic(V) and antimony(V) halides," *J. Chem. Soc. A*, 2166–8 (1968).

J. R. Durig and J. S. DiYorio, "Vibrational spectra of organophosphorus compounds. IV. Infrared and Raman spectra of CH_3OPCl_2 and CD_3OPCl_2," *J. Chem. Phys.*, **48**, 4154–61 (1968).

R. A. Nyquist, "Vibrational spectroscopic study of $(RPO_3)^{2-}$," *J. Mol. Struct.*, **2**, 123–35 (1968).

I. R. Beattie, K. Livingstone and T. Gilson, "The vibrational spectrum of methyltetrachlorophosphorous (V)," *J. Chem. Soc. A*, 1–3 (1968).

J. R. Durig, D. W. Wertz, B. R. Mitchell, F. Block, and J. M. Greene, "Vibrational spectra of organophosphorus compounds. III. Infrared and Raman spectra of $(CH_3)_2PSCl$, $(CH_3)_2PSBr$ and $(CH_3)_2POCl$," *J. Phys. Chem.*, **71**, 3815–23 (1967).

P. J. Hendra and L. Jovic, "Infrared and Raman spectra of some complexes between thiourea and tellurium (II). II," *J Chem. Soc., A*, 911–13 (1968).

R. A. Walton, "Coordination compounds of thallium (III). V. The vibrational spectra of mixed tetrahalothallates of the type $Et_4-NTlX_{4-x}Y_{4x}$, where $x = 1, 2$, or 3," *Inorg. Chem.*, **7**, 1927–30 (1968).

R. S. Tobias, M. J. Sprague, and G. E. Glass, "Reactions of dimethylgallium (III) hydroxide. Raman, infrared, and proton magnetic resonance spectra of the dimethylgallium (III) aquo ion and several of its compounds," *Inorg. Chem.*, **7**, 1714–21 (1968).

G. A. Lee, "Raman spectrum of tris (trimethylsilyl) thallium," *Spectrochim. Acta*, **25A**, 1841 (1969).

E. V. Van den Berghe, L. Verdonck, and G. P. Van der Kelen, "Ethyltin chloridepyridine complexes," *J. Organometal. Chem.*, **16**, 497–9 (1969).

V. A. Maroni and T. G. Spiro, "Raman and infrared spectra of tetrameric thallium (I) alkoxides and their analysis," *Inorg. Chem.*, **7**, 193–7 (1968).

R. J. Capwell, K. H. Rhee, and K. S. Seshadri, "Vibrational spectra of sodium and lithium methane-sulfonates," *Spectrochim Acta.*, **24A**, 955–8 (1968).

J. R. Durig, C. W. Sink, and J. B. Turner, "Vibrational spectra and structure of organic germanes. Normal vibrations and free rotation in phenylgermane," *J. Chem. Phys.*, **49**, 3422–41 (1968).

J. R. Durig and J. S. DiYorio "Vibrational spectra of organophosphorus compounds. Infrared and Raman spectra of methylphosphorylchloride and methyl-O (3)-phosphorylchloride," *J. Chem. Phys.*, **48**, 4154–61 (1968).

J. R. Durig and J. S. DiYorio, "Vibrational spectra of organo phosphorus compounds. Tetramethyl bi-phosphine," *Inorg. Chem.*, **8**, 2796–802 (1969).

A. L. Smith, "Low frequency vibrational spectra of group IVA phenyl compounds," *Spectrochim. Acta.*, **24A**, 695–706 (1968).

W. H. Green and A. B. Harvey, "Vibrational spectra and structure of dimethyldiselenide and dimethyldiselenide-d_6," *J. Chem. Phys.*, **49**, 3586–95 (1968).

ORGANIC COMPOUNDS—ALIPHATIC

N. K. Sanyal, A. N. Pandey, and H. S. Singh, "Correlation between the Raman frequencies and boiling points of normal and substituted hydrocarbons," *J. Quant. Spectrosc. Radiat. Transfer.*, **9**, 465–8 (1969).

C. M. Pathak and W. H. Fletcher, "Infrared and Raman spectra of isobutene and isobutene-d_8," *J. Mol. Spectrosc.*, **31**, 32–53 (1969).

A. D. H. Clagni and A. Danti, "Low frequency infrared and Raman spectra of certain aliphatic ethers," *Spectrochim. Acta*, **24A**, 439–41 (1968).

H. Wieser, W. G. Laidlaw, P. J. Krueger, and H. Fuhrer, "Vibrational spectra and a valence force field for conformers of diethyl ether and deuterated analogs," *Spectrochim. Acta*, **24A**, 1055–89 (1968).

S. M. Blumenfeld and H. Fast, "Low-frequency Raman spectra of solid and liquid formic acid," *Spectrochim. Acta*, **24A**, 1449–59 (1968).

M. Suzuki and T. Shimanouchi, "Infrared and Raman spectra of succinic acid crystals," *J. Mol. Spectrosc.*, **28**, 394–410 (1968).

M. Suzuki and T. Shimanouchi, "Infrared and Raman spectra of adipic acid crystal," *J. Mol. Spectrosc.*, **29**, 415–25 (1969).

J. Maillols, L. Bardet, and R. Marignan, "Use of the Raman effect to study the structure of potassium hydrogen maleate. I. The hydrogen maleate ion in solution," *J. Chim. Phys. Physicochim. Biol.*, **66**, 522–8 (1969). (Fr.)

J. Maillols, L. Bardet, and R. Marignan, "Use of the Raman effect to study the structure of potassium hydrogen maleate. II. Single-crystal study. Internal vibrations," *J. Chim. Phys. Physicochim. Biol.*, **66**, 529–38 (1969) (Fr.).

J. Maillols, L. Bardet, and R. Marignan, "Use of the Raman effect to study the structure of potassium hydrogen maleate. III. Single-crystal study. Low-frequency spectra," *J. Chim. Phys. Physicochim. Biol.*, **66**, 539–47 (1969) (Fr.).

A. M. Bellocq, R. Martegoutes, M. Dubien, J. Belloc, P. Dizabo, and C. Garrigou-Lagrange, "Vibrational spectra of isobutyric acid, deuterated isobutyric acid, and their potassium and silver salts," *J. Chim. Phys. Physicochim. Biol.*, **66**, 449–58 (1969) (Fr.).

N. O. George and F. V. Robinson, "Vibrational spectra of sodium acetylacetonate and sodium acetylacetnoate-d," *J. Chem. Soc.*, A, 1950–4 (1968).

R. J. Capwell, Jr., "Infrared and Raman spectra of acetaldehyde-2,2,2,-d_3," *J. Chem. Phys.*, **49**, 1436–8 (1968).

H. Michelsen and P. Klaboe, "Spectroscopic studies of glycolaldehyde," *J. Mol. Struct.*, **4**, 293–302 (1969).

R. S. Dennen, E. A. Piotrowski, A. Edward, and F. F. Cleveland, "Raman and infrared spectral data for dibromomethane, dibromomethane-d, and dibromomethane-d_2," *J. Chem. Phys.*, **49**, 4385–91 (1968).

R. P. Fournier, R. Savoie, F. Bessette, and A. Cabana, "Vibrational Spectra of liquid and crystalline carbon tetrafluoride," *J. Chem. Phys.*, **49**, 1159–64 (1968).

M. C. Hutley and D. J. Jacobs, "Angular dependence of Raman scattering in carbon tetrachloride," *Opt. Commun.*, **1**, 167–8 (1969).

C. Altona and H. J. Hageman, "Conformation of open-chain compounds. III. Carbon-halogen stretching frequencies and conformation of some vicinal dihalides," *Recl. Trav. Chim. Pays-Bas*, **88**, 33–42 (1969).

F. A. Miller and F. E. Kiviat, "Vibrational spectra of several ethyl chlorides $CH_3CH_2Cl, CH_3CD_2Cl, CD_3CH_2Cl$ and CD_3CD_2Cl," *Spectrochim. Acta*, **25A**, 1363–74 (1969).

N. C. Craig and J. Overend, "Vibrational assignments and potential constants for cis-and trans-1,2, difluoroethylenes and their deuterated modifications," *J. Chem. Phys.*, **51**, 1127–42 (1969).

E. J. Flourie and W. D. Jones, "Vibrational spectra and assignments of tetraiodoethylene," *Spectrochim. Acta*, **25A**, 653–9 (1969).

R. Forneris and D. Bassi, "Vibrational spectra and force constants of tetraiodoethylene," *J. Mol. Spectrosc.*, **26**, 220–6 (1968).

P. Klaeboe and E. Kloster-Jense, "Raman spectra and revised vibrational assignments of some halocyanoacetylenes," *Spectrochim. Acta*, **23A**, 1981–90 (1967).

P. J. D. Park and E. Wyn-Jones, "Infrared and Raman studies on meso- and (I)-2,3-dichloro- and -2,3-dibromobutanes," *J. Chem. Soc. A*, 422–6 (1969).

P. Glyzinski and Z. Eakstein, "Vibrational spectra of α-halonitroalkanes. I. Monohalonitromethanes and their deuterated derivatives," *Spectrochim Acta*, **24A**, 1777–84 (1968).

F. A. Miller and F. E. Kiviat "Infrared and Raman spectra of $(CF_3)_2C:C:O$, $(CF_3)_2C:N:N$, and $(CF_3)_2C:NH$," *Spectrochim. Acta*, **25A**, 1577–88 (1969).

G. Lucazeau and A. Novak "Vibrational spectra of the aldehydes CCl_3CHO, CCl_3CDO and CBr_3CHO," *Spectrochim. Acta*, **25A**, 1615–29 (1969) (Fr.).

J. R. Durig and D. W. Wertz, "The infrared and Raman spectra of bis(trifluoromethyl.)-peroxide," *J. Mol. Spectrosc.*, **25**, 465–76 (1968).

R. A. Nyquist, "Vibrational spectroscopic study of methyl thiolchloroformate," *J. Mol. Struct.*, **1**, 1–10 (1967).

K. Sathiavandan and J. L. Morgrave, "Vibrational spectra of (bis(perfluoroisopropyl)) sulfur difluoride and (trifluoromethyl) sulfur trifluoride," *Ind. J. Pure Appl. Phys.*, **5**, 464–7 (1967).

D. A. Bahnick and W. B. Person, "Raman intensity study of charge-transfer complexes of iodine cyanide," *J. Chem. Phys.*, **48**, 5637–45 (1968).

M. G. Miles, G. Doyle, R. P. Cooney, and R. S. Tobias, "Raman and infrared spectra and normal coordinates of the trifluoromethanesulfonate and trichloromethanesulfonate anions," *Spectrochim. Acta*, **25A**, 1515–26 (1969).

D. H. Christensen, N. Rastrup-Anderson, D. Jones, P. Klaeboe and E. R. Lippincott, "Infrared, Raman and proton magnetic resonance spectra of methyl thionitrite," *Spectrochim. Acta*, **24A**, 1581–9 (1968).

S. G. Frankiss, "Vibrational spectra and structures of $S_2(CH_3)_2$ and $Se_2(CH_3)_2$," *J. Mol. Struct.*, **3**, 88–101 (1969).

J. Fabian, H. Krocber, and R. Mayer, "Vibrational spectroscopic studies on vinyl methyl sulfide," *Spectrochim. Acta*, **24A**, 727–36 (1968) (Ger.).

P. Klaeboe, "Vibrational spectra of dimethyl sulfite," *Acta, Chem. Scand.*, **22**, 2817–21 (1968).

K. Herzog, E. Steger, P. Rosmus, S. Scheithauer, and R. Mayer, "Infrared and Raman spectra of methyl dithioacetate and dimethyl trithiocarbonate," *J. Mol. Struct.*, **3**, 339–50 (1969) (Ger.).

P. Klaboe "Vibrational spectra of 1,4-dithiane and 1.3.5-trithiane," *Spectrochim. Acta*, **25A**, 1427–47 (1969).

J. H. S. Green, D. J. Harrison, W. Kynaston, and D. W. Scott, "Vibrational spectrum of 2,2,5,5-tetramethyl-3,4-dithiahexane," *Spectrochim. Acta*, **25A**, 1313–14 (1969).

J. A. Creighton, J. H. S. Green, and D. J. Harrison, "Vibrational spectra of some trimethylsulfoxonium salts, d_9, and h_9," *Spectrochim. Acta*, **25A**, 1314–18 (1969).

A. Blaschette and H. Buerger, "Vibrational spectra of some trimethyl-h (and d_9)-sulfoxonium halides," *Inorg. Nucl. Chem. Lett.*, **5**, 639–42 (1969) (Ger.).

J. A. Creighton, J. H. S. Green, D. J. Harrison, and S. M. Waller, "The vibrational spectra of some trimethylsulfonium and trimethylsulfoxonium compounds," *Spectrochim. Acta*, **23A**, 2973–9 (1967).

R. W. Mitchell, F. A. Hartman, and J. A. Merritt, "Infrared, Raman and NMR spectra of ethylene episulfoxide," *J. Mol. Spectrosc.*, **31**, 388–99 (1969).

D. H. Christensen and D. F. Nielson, "Infrared and Raman Spectra of 1,2,5-oxadiazole. Vibrational assignment and thermodynamic functions of 1,2,5-oxadiazole," *J. Mol. Spectrosc.*, **24**, 477–89 (1967).

M. J. Brookes and N. Jonathan, "Spectroscopic evidence for the existence of geometric isomers of the sodium salts of nitroparaffins," *Spectrochim. Acta*, **25A**, 187–91 (1969).

J. R. Durig, S. F. Bush, and F. G. Baglin, "Infrared and Raman investigation of condensed phases of methylamine and its deuterium derivatives," *J. Chem. Phys.*, **49**, 2106–17 (1968).

P. H. Clippard and R. C. Taylor, "Raman spectra and vibrational assignments for trimethylamine," *J. Chem. Phys.*, **50**, 1472–3 (1969).

P. Klaeboe and J. Grundnes, "Vibrational spectra of propionitrile, 2-chloro-, and 2-bromoproprionitrile," *Spectrochim. Acta*, **24A**, 1905–16 (1968).

M. F. ElBermani and N. Jonathan, "Spectroscopic studies of rotational isomerism in β-halopropionitriles," *J. Chem. Soc. A*, 1711–14 (1968).

L. A. Franks, A. J. Merer, and K. K. Innes, "Analysis of the infrared and Raman spectra of tetrazine-*d* & *d₂*," *J. Mol. Spectrosc.*, **26**, 458–64 (1968).

J. R. Durig and J. W. Clark, "Raman spectrum and structure of tetrafluorohydrazine," *J. Chem. Phys.*, **48**, 3216–25 (1968).

J. R. Durig, W. C. Harris, and D. W. Wertz, "Infrared and Raman spectra of substituted hydrazines. I. Methylhydrazine," *J. Chem. Phys.*, **50**, 1449–61 (1969).

P. H. Mogril, "Physical and optical properties of carbodiimides," *Nucl. Sci. Abstr.*, **21**, 470–14 (1967).

M. Harrand and C. Fegly, "Raman spectrographic study of cis-trans isomerism in maleic and fumaric acids," *J. Chim. Phys.*, **64**. 991–6 (1967) (Fr.).

C. V. Berney, "Spectroscopy of trifluoroacetaldehyde, trifluoroacetylfluoride and trifluoroacetylchloride. Vibrational spectrum and barrier to internal rotation of trifluoroacetaldehyde," *Spectrochim. Acta*, **25A**, 793–809 (1969).

J. R. Durig and W. C. Harris, "Infrared and Raman spectra of substituted hydrazines. Unsymmetrical dimethyl hydrazine," *J. Chem. Phys.*, **51**, 4457–68 (1969).

C. W. Brown, R. J. Obremski, J. R. Allkins, and E. R. Lippincott, "Vibrational spectra of single crystals and polycrystalline films of dichloromethane and dibromethane," *J. Chem. Phys.*, **51**, 1376–84 (1969).

R. D. McLachlan and R. A. Nyquist, "Vibrational spectra of the allyl halides," *Spectrochim. Acta*, **24A**, 103–114 (1968).

J. R. Durig and D. W. Wertz, "Far infrared and Raman spectra of the haloacetonitriles," *Spectrochim. Acta*, **24A**, 21–9 (1968).

ORGANIC COMPOUNDS—ALIPHATIC RINGS

C. Altona, "Structure-frequency correlations and conformation of alkyl and cycloalkyl halides," *Tetrahedron Lett.*, 2325–30 (1968).

V. Schettino, M. P. Marzocchi, and G. Sbrana, "Vibrational crystal spectrum of 2,2,7-paracyclophane," *J. Mol. Struct.*, **2**, 39–45 (1968).

J. B. Bates, D. E. Sands, and W. H. Smith, "Spectra and structure of crystalline cyclopropane and cyclopropane-d_6," *J. Chem. Phys.*, **51**, 105–113 (1969).

R. W. Mitchell, E. A. Dorko, and J. A. Merritt, "Vibrational spectra of cyclopropene and cyclopropene-1,2-d$_2$," *J. Mol. Spectrosc.*, **26**, 197–212 (1968).

F. A. Miller, F. E. Kiviat, and J. Matsufara, "Infrared and Raman spectra of cyclobutane-1,3-dione," *Spectrochim. Acta*, **24A**, 1523–6 (1968).

J. R. Durig, J. M. Karriker, and D. W. Wertz, "Vibrational spectra and structure of small ring compounds, Chlorocyclopentane," *J. Mol. Spectrosc.*, **31**, 237—55 (1969).

E. Gallenella, B. Fortunato, and P. Mirone, "Infrared and Raman spectra and vibrational assignments of cyclopentadiene," *J. Mol. Spectrosc.*, **24**, 345–62 (1967).

R. J. Obremski, C. W. Brown, and E. R. Lippincott, "Vibrational spectra of single crystals. Polymorphic solids of cyclohexane," *J. Chem. Phys.*, **49**, 185–91 (1968).

C. DiLauro, N. Neto, and S. Califano, "Vibrational spectrum and normal-mode analysis of 1,3-cyclohexadiene," *J. Mol. Struct.*, **3**, 219–26 (1969).

G. R. Olah, A. Commeyras, and U. Y. Lui, "Stable carbonium ions. LXXXII. Raman and N.M.R. spectroscopic study of the nortricyclonium ion (protonated tricyclo (2.2.1.0^2,6) heptane) and its relation to the 2-norbornyl (bicyclo (2.2.1)heptyl) cation. The nature of the stable, long-lived norbornyl cation in strong acid solutions," *J. Amer. Chem. Soc.*, **90**, 3882–4 (1968).

N. Neto, C. DiLauro, and S. Califano, "Infrared, Raman spectrum, and normal coordinate calculation of *trans, trans, trans*-cyclodecatriene," *Spectrochim. Acta*, **24A**, 385–93.

J. R. Durig, S. F. Bush, and W. C. Harris, "Vibrational spectra and structure of small-ring compounds. XII. Propyleneimine (2-methylaziridine) and propyleneimine-d$_1$," *J. Chem. Phys.*, **50**, 2851–63 (1969).

J. R. Durig and A. C. Morissey, "Vibrational spectra and structure of small-ring compounds. XI. β-Butyrolactone," *J. Mol. Struct.*, **2**, 377–90 (1968).

J. R. Durig and W. H. Green, "Vibrational spectra and structure of four-membered ring molecules. IV. 2-Bromocyclobutanone and 2-bromo-2,4,4-trideuterocyclotubanone," *J. Chem. Phys.*, **47**, 4455–62 (1967).

J. R. Durig, G. L. Coulter, and D. W. Wertz, "Far-infrared spectra and structure of small ring compounds. Ethylene carbonate, γ-butyrolactone, and cyclopentanone," *J. Mol. Spectrosc.*, **27**, 285–95 (1968).

ORGANIC COMPOUNDS—HETEROCYCLICS

E. Castelucci, G. Sbrana, and F. D. Verderame, "Infrared spectra of crystalline and matrix isolated pyridine and pyridine-d$_5$," *J. Chem. Phys.*, **51**, 3762–70 (1969).

R. T. Bailey and D. Steele, "The vibrational spectra of substituted nitrogen heterocyclic systems. II. 2,6-Difluoropyridine," *Spectrochim. Acta*, **23A**, 2997–3005 (1967).

S. G. W. Ginn, I. Haque, and H. L. Wood "Vibration spectra of the complexes pyridene-bromine and pyridine-brominechloride," *Spectrochim. Acta*, **24A**, 1531–42 (1968).

R. Foglizzo and A. Novak, "Low-frequency infrared and Raman spectra of hydrogen-bonded pyridinium halides," *J. Chem. Phys.*, **50**, 5366–73 (1969).

R. Foglizzo and A. Novak, "Raman spectrum of crystalline pyrimidene," *Spectrosc. Lett.*, **2**, 165–71 (1969).

R. T. Bailey and D. Steele, "The vibrational spectra of substituted nitrogen heterocyclic systems. I. 2,4,6-Trifluoropyrimidine," *Spectrochim. Acta*, **23A**, 2989–95
(1967).

R. T. Bailey and D. Steele, "Vibrational spectra of substituted nitrogen heterocyclic
systems. III. 1,3,5-trichloropyrimidine," *Spectrochim. Acta*, **25A**, 219–25 (1969).

L. Colombo, "Low-frequency Raman spectrum of imidazole single crystals," *J. Chem.
Phys.*, **49**, 4688–95 (1968).

M. Rico, M. Barrachina, and J. M. Orza, "Fundamental vibrations of furan and
deuterated derivatives," *J, Mol. Spectrosc.*, **25**, 133–48 (1967).

C. P. Giryarallablan and K. Venkateswash, "Raman spectrum of coumarin,"
Curr. Sci., **37**, 10–11 (1968).

W. H. Green and A. B. Harvey, "Vibrational spectra of 2,5-dihydrothiophene."
Spectrochim. Acta, **25A**, 723–30 (1969).

M. Cordes and J. L. Water, "Infrared and Raman spectra of heterocyclic compounds.
Infrared studies and normal vibrations of imidazole," *Spectrochim. Acta*, **24A**, 237–
52 (1968).

J. L. Walter and M. M. Cordes, "Infrared and Raman studies of heterocyclic compounds. Infrared spectra and normal vibrations of benzimidazole and bis
(benzimidazolato) metal complexes," *Spectrochim. Acta*, **24A**, 1421–35 (1968).

A. Bree and R. Zwarick, "Vibrational assignment of carbozole from infrared, Raman
and fluorescence spectra," *J. Chem. Phys.*, **49**, 3334–55 (1968).

W. H. Green and A. B. Harvey, "Vibrational spectra of 2,5-dihydro thiophene,"
Spectrochim. Acta, **25A**, 723—30 (1969).

R. Foglizzo and A. Novak, "Low frequency infrared and Raman spectra of hydrogen
bonded pyridinium halides," *J. Chem. Phys.*, **50**, 5366–73 (1969).

A. B. Harvey, J. R. Durig, and A. C. Morrisey, "Vibrational spectra and structure
of four membered ring molecules. Vibrational analysis and ring puckering
vibration of trimethylselenide and trimethylselenide-d_6," *J. Chem. Phys.*, **50**,
4949–61 (1969).

ORGANIC COMPOUNDS—AROMATICS

W. G. Fateley, G. L. Carlson, and F. E. Dickson, "Infrared and Raman studies of
ortho- and meta-substituted styrenes," *Appl. Spectros.*, **22**, 650–8 (1968).

J. E. Saunders, J. J. Lucier, and J. N. Willis, Jr., "Infrared and Raman vibrational
study of ethylbenzene, n-propylbenzene and terminal halogen-substituted analogs, 800–50 cm^{-1}," *Spectrochim. Acta*, **24A**, 2023–43 (1968.)

P. N. Gates, K. Radcliffe, and D. Steele, "Vibrational spectra of mono- and para-
disubstituted halobenzenes," *Spectrochim. Acta*, **25A**, 507–16 (1969).

M. Suzuki and M. Ito, "Polarized Raman spectra of *p*-dichlorobenzene and *p*-
dibromobenzene single crystals," *Spectrochim. Acta*, **25A**, 1017–21 (1969).

A. H. Hikal, W. Wolf, and A. B. Burg, "Far infrared and Raman spectra of aryliodo-
dichlorides," *Spectrosc. Lett.*, **2**, 13–18 (1969).

R. T. Bailey and S. G. Hasson, "Vibrational spectra of pentafluorotoluene," *Spectrochim. Acta*, **25A**, 467–73 (1969).

R. T. Bailey and S. G. Hasson, "Vibrational spectra of perfluorotoluene," *Spectrochim. Acta*, **24A**, 1891–8 (1968).

H. F. Sharvell, A. S. Blair, and R. J. Jakobsen, "Infrared and Raman spectra of
pentafluorobenzonitrile," *Spectrochim. Acta*, **24A**, 1257–66 (1968).

G. Zerbi and S. Sandroni, "Fundamental frequencies and molecular configuration of biphenyl. I. Re-analysis of its vibrational spectrum," *Spectrochim. Acta*, **24A**, 483–510 (1968).

S. A. Solin and A. K. Ramdas, "Raman spectrum of crystalline benzil," *Phys. Rev.*, **174**, 1069–75 (1968).

J. Brandmueller and R. Claus, "Intensity measurements on the crystal lattice vibrations of anthracene and naphthalene," *Spectrochim. Acta*, **25A**, 103–10 (1969) (Ger.).

A. Bree and R. A. Kydd, "Raman spectrum of anthracene-d_{10}," *Chem. Phys. Lett.*, **3**, 357–60 (1969).

A. Bree, R. A. Kydd, and T. N. Misra, "Vibrational assignment of acenaphthene," *Spectrochim. Acta*, **25A**, 1815–29 (1969).

H. F. Shurvell, A. R. Norris, and D. E. Irish, "Raman and far infrared spectra of *s*-trinitrobenzene and *s*-trinitiobenzene-d_3," *Can. J. Chem.*, **47**, 2515–19 (1969).

McHulby and D. J. Jacobs, "Resonance Raman effect in *p*-nitroaniline," *Chem. Phys. Lett.*, **3**, 711–14 (1969).

H. Lee and J. K. Wilmshurst, "Infrared and Raman spectra of phenyl acetate, phenyl-d_5 acetate, and phenyl acetate-d_3," *Aust. J. Chem.*, **22**, 691–700 (1969).

F. Stenman "Raman scattering from powdered benzil," *J. Chem. Phys.*, **51**, 3141–3 (1969).

S. C. Sirkar and P. K. Bishui, "Raman and infrared spectra of benzoyl chloride in different states," *Indian J. Phys.*, **42**, 243–53 (1968).

L. Bardet, G. Fleury, R. Granger, and C. Sablayrolles, "Structural study of indane and indanones. II. Structure of 2-indanone," *J. Mol. Struct.*, **3**, 129–39 (1969) (Fr.).

L. Bardet, G. Fleury, R. Granger, and C. Sablayrolles, "Structural study of indane and indanones. I. Structure of indane," *J. Mol. Struct.*, **2**, 397–408 (1968) (Fr.).

A. Perrier-Datin and J. M. Lebas, "Comparative vibrational study of diphenyl methylenimine, *N*-deuterodiphenylmethylenimine, and benzophenone in solution," *Spectrochim. Acta.*, **25A**, 169–85 (1969) (Fr.).

G. A. Olah, J. R. DeMember, C. W. Lui, and A. M. White, "Stable carbonium ions, hydrogen-1 and carbon-13 nuclear magnetic resonance and laser Raman spectroscopic study of the 2-methyl,-2-ethyl and 2-phenylnorbornylcations," *J. Amer. Chem. Soc.*, **91**, 3958–60 (1969).

J. H. S. Green, "Vibrational spectra of benzene derivatives. IV. Methylphenyl sulfide, diphenyl sulfide, diphenyl disulfide, and diphenyl sulfoxide," *Spectrochim. Acta*, **24A**, 1627–37 (1968).

INORGANIC COMPOUNDS—GENERAL

V. A. Maconi and T. G. Spiro, "Vibrational analysis for polynuclear hydroxylead (II) complexes," *Inorg. Chem.*, **7**, 188–92 (1968).

W. G. Nilsen and J. G. Skinner, "Raman spectrum of strontium titanate," *J. Chem. Phys.*, **48**, 2240–8 (1968).

F. A. Grimm, L. Barton, and R. F. Power, "Vibrational analysis of gaseous boroxine," *Inorg. Chem.*, **7**, 1309–15 (1968).

J. R. Durig and K. L. Hellams, "Vibrational spectra and structure of some silicon containing compounds. Hexachlorodisiloxane," *Inorg. Chem.*, **8**, 944–50 (1969).

O. Oehler and H. H. Guenthard, "Calcium hydroxide and calcium deuterated hydroxide-vibrational crystal spectra, normal coordinate analysis and assignment," *J. Chem. Phys.*, **48**, 2036–45 (1968).

J. E. Newberry, "Vibrational spectra of the dioxotetrachloro uranate (VI) and the dioxotetrabromo uranate (VI) ions," *Spectrochim. Acta*, **25A**, 1699–702 (1969).

F. A. Miller and D. H. Lemmon, "Infrared and Raman spectra of trichloro mercapto silane and the partial spectra of trichlorocyano silane," *Spectrochim. Acta*, **25A**, 1799–806 (1969).

ALKALI METAL COMPOUNDS

B. M. Chadwick and S. G. Frankiss, "Vibrational spectra and structure of poly-crystalline $KAg(CN)_2$, $NaAg(CN)_2$ and $TlAg(CN)$," *J. Mol. Struct.*, **2**, 281–5 (1968).

G. L. Bottger, "Vibrational spectrum of crystalline $KAg(CN)_2$," *Spectrochim. Acta*, **24A**, 1821–9 (1968).

R. E. Miller, R. R. Getty, K. L. Treuil, and G. E. Leroi, "Raman spectrum of crystalline lithium nitrate," *U. S. Govt. Res. Develop. Rep.*, **68**, 67 (1968).

W. H. Leong and D. W. James, "Vibrational spectra of anhydrous lithium per-chlorate in crystalline and molten states," *Aust. J. Chem.*, **22**, 499–503 (1969).

D. W. James and W. H. Leong, "Structure of molten nitrates: temperature variation of Raman spectrum of sodium nitrate," *Chem. Commun.*, 1415–17 (1968).

D. W. James and W. H. Leong, "Vibrational spectra of single crystals of Group I nitrates," *J. Chem. Phys.*, **49**, 5089–96 (1969).

M. Balkanski, M. Teng, and M. Nusimovici, "Raman scattering in potassium nitrate phases I, II, and III," *Phys. Rev.*, **176**, 1098–1106 (1968).

E. V. Chisler, "Raman spectrum study of a phase transition in a sodium nitrate crystal," *Fiz. Tverd. Tela.*, **11**, 1272–81 (1969) (Russ).

R. E. Miller, R. R. Getty, K. L. Treuil, and G. E. Leroi, "Raman spectrum of crystalline lithium nitrate," *J. Chem. Phys.*, **51**, 1385–9 (1969).

D. L. Rousseau, R. E. Miller, and G. E. Leroi, "Raman spectrum of crystalline sodium nitrate," *J. Chem. Phys.*, **48**, 3409–13 (1968).

I. Nakagawa and J. L. Walter "Optically active crystal vibrations of the alkali-metal nitrates," *J. Chem. Phys.*, **51**, 1389–97 (1969).

M. Tsuboi, M. Terada, and T. Kajiura, "Raman effect in ferroelectric sodium nitrate crystal. II. Dependence of B_2 type phonon spectrum on propagation direction," *Bull. Chem. Soc., Japan*, **42**, 1871–4 (1969).

P. Li and J. P. Devlin, "Vibrational spectra and structures of ionic liquids. Potassium nitrate-silver nitrate mixtures," *J. Chem. Phys.*, **49**, 1441–2 (1968).

K. Williamson, P. Li, and J. P. Devlin, "Vibrational spectra and structures of ionic liquids. II. Pure alkali metal nitrates," *J. Chem. Phys.*, **48**, 3891–6 (1968).

ALUMINUM AND INDIUM

M. Ashkin, J. H. Parker, Jr., and D. D. Feldman, "Temperature dependence of the Raman lines of α-aluminum oxide," *Solid State Commun.*, **6**, 363–6 (1968).

J. F. Scott, "Raman study of trigonal-cubic phase transistions in rare earth alumin-ates," *Phys. Rev.*, **183**, 823–5 (1969).

J. A. Koningstein, "Comparison of far-infrared and Raman spectra. of some rare earth garnet single crystals." *Chem. Phys. Lett.* **3**, 303–4 (1969).

I. R. Beattie, T. Gilson, and G. A. Ozin, "Vibrational spectra of Al_2Br_2, Al_2I_6, Ga_2Br_6, Ga_2I_6 and In_2I_6," *J. Chem. Soc. A*, 813–15 (1968).

D. M. Adams and R. G. Churchill, "Vibrational spectra of halogen-bridged systems. II. Au_2Cl_6, Al_2Br_6, Al_2I_6, and In_2I_6," *J. Chem. Soc. A*, 2141–4 (1968).

O. Brafman, G. Lengyel, S. S. Mitra, P. J. Gielisse, J. N. Phendl, and L. C. Mansur, "Raman spectra of aluminum nitrate, cubic boron nitride, and boron phosphide," *Solid State Commun.*, **6**, 523–6 (1968).

J. H. R. Clarke and R. E. Hester, "Raman spectroscopic studies of the structure of the lower chlorides of indium in the molten state." *Inorg. Chem.*, **5**, 113–16 (1969).

J. H. R. Clarke and R. E. Hester, "Raman spectrum and structure of molten indium dichloride," *Chem. Commun.*, **17**, 1042–3 (1968).

J. H. R. Clarke and R. E. Hester, "Vibrational spectra of molten salts. VI. Raman spectra of indium trichloride—alkali metal chloride mixture," *J. Chem. Phys.*, **50**, 3106–12 (1969).

N. N. Greenwood, D. J. Prince, and B. P. Straughan, "Vibrational spectra and structures of indium (III) chloride, bromide, and iodide," *J. Chem. Soc. A*, 1694–6 (1968).

D. F. Shriver and I. Wharf, "Preparations, far infrared spectra and Raman spectra of pentahalo indium (III) and thallium (III) complexes." *Inorg. Chem.*, **8**, 2167–71 (1969).

J. H. R. Clarke and R. E. Hester, "Raman spectroscopic studies of the structure of the lower chlorides of indium in the molten state," *Inorg. Chem.*, **8**, 1113–16 (1969).

T. Ogawa, "Vibrational assignments and normal vibrations of trimethylaluminum," *Spectrochim. Acta*, **24A**, 15–20 (1968).

ARSENIC AND ANTIMONY

H. A. Szymanski, L. Marabella, J. Hoke, and J. Harter, "Infrared and Raman studies of arsenic compounds," *Appl. Spectrosc.*, **227**, 290–304 (1968).

J. E. Drake and J. Simpson, "Infrared and Raman spectra of monosilylarsine," *Spectrochim. Acta*, **24A**, 981–4 (1968).

J. E. D. Davies and D. A. Long, "Vibrational spectra of halides and complex halides of arsenic (III). I. Liquid and solid arsenic trichloride and arsenic tribromide, and their solutions in tributyl phosphate," *J. Chem. Soc. A*, 1757–60 (1968).

J. E. D. Davies and D. A. Long, "The vibrational spectra of halides and complex halides of arsenic (III). II. The tetrachloroarsenate (III) and tetrabromoarsenate (III) anions," *J. Chem. Soc. A*, 1761–5 (1968).

M. A. Hooper and D. W. James, "Vibrational spectra of arsenic triodide," *Aust. J. Chem.*, **21**, 2379–83 (1968).

J. A. Evans and D. A. Long, "Vibrational Raman spectra of some complexes formed between the tetrafluourides of Group VI elements and antimony and arsenic pentafluorides," *J. Chem. Soc. A*, 1688–94 (1968).

P. K. Bishui and S. C. Sirkar, "Raman spectrum of antimony trichloride at $-195°$," *Indian J. Phys.*, **42**, 330–2 (1968).

K. Olie, C. C. Smitskamp, and H. Gerding, "Two modifications of solid antimony pentachloride," *Inorg. Nucl. Chem. Lett.*, **4**, 129–32 (1968).

M. A. Hooper and D. W. James, "Raman spectra of antimony triiodide," *Spectrochim. Acta*, **25A**, 569–70 (1969).

R. A. Walton, "Complex halides of nontransition metals. II. The vibrational spectra of mono and dibutylammonium salts of the $SbCl_6{}^{3-}$, $BiCl_6{}^{3-}$, $BiBr_6{}^{3-}$, $BiCl_5{}^{2-}$, and $SbBr_5{}^{2-}$ anions," *Spectrochim. Acta*, **24A**, 1527–30 (1968).

D. C. McKean, "Spectroscopic evidence for geometry in trisilylphosphine and trisilylarsine," *Spectrochim. Acta*, **24A**, 1252–4 (1968).

R. D. Feltham and W. Silverthorn, "Spectra of bis(tertiaryarsine) complexes," *Inorg. Chem.*, **7**, 1154–8 (1968).

BISMUTH

R. P. Oertel and R. A. Plane, "Raman study of chloride and bromide complexes of bismuth (III)," *Inorg. Chem.*, **6**, 1960–7 (1967).

R. P. Oertel and R. A. Plane, "Raman and infrared study of nitrate complexes of bismuth (III)," *Inorg. Chem.*, **7**, 1192–6 (1968).

I. R. Beattie, K. M. S. Livingston, G. A. Ozin, and I. J. Reynolds, "Single-crystal Raman spectrum of bismuth pentafluoride and of antimony tetrachloridefluoride and the vibrational spectrum of vanadium, niobium, tantalum, and antimony pentafluorides," *J. Chem. Soc. A*, 958–65 (1969).

R. A. Spragg, H. Stammreich, and Y. Kawano, "Raman spectra and structure of bismuth iodide complexes," *J. Mol. Struct.*, **3**, 305–9 (1969).

R. P. Oertel and R. A. Plane, "Raman spectra of solid bismuth (III) bromide and chloride," *Inorg. Chem.*, **8**, 1188–90 (1969).

J. T. Kenney and F. X. Powell, "Raman spectra of fused indium and bismuth chlorides," *J. Phys. Chem.*, **72**, 3094—7 (1968).

V. A. Maconi and T. G. Spiro, "Normal coordinate analysis for dodecahydroxydecabismuth $(6+)$ ion. Evidence for bismuth-bismuth bonding," *Inorg. Chem.*, **7**, 183–7 (1968).

R. P. Oertel and R. A. Plane, "Raman and infrared study of nitrate complexes of bismuth (III)," *Inorg. Chem.*, **7**, 1192–6 (1968).

CADMIUM, CHROMIUM, AND COBALT

M. S. Mathur, C. A. Frenzel, and E. B. Bradley, "Measurements of the Raman and the infrared spectrum of potassium dichromate," *J. Mol. Struct.*, **2**, 429–35 (1968).

J. E. D. Davies and D. A. Long, "Vibrational spectra of the halides and complex halides of the Group IIB elements. I. The vibrational spectra of CdI_3, $CdBr_2$, $CdBr_3$, $CdCl_2$, $CdCl_3$, $CdCl_4{}^{2-}$ and $MgCl_4{}^{2-}$," *J. Chem. Soc. A*, 2054–8 (1968).

D. A. Brown, D. Cunningham, and W. K. Blass, "The infrared and Raman spectra of chromium (III) oxide," *Spectrochim. Acta*, **24A**, 965–8 (1968).

R. H. Mann, I. J. Hyams, and E. R. Lippincott, "Laser Raman spectrum and force constant calculation of $Co(CO)_3NO$," *J. Chem. Phys.*, **48**, 4929–34 (1968).

G. Harbeke and E. J. Steigmeier, "Raman scattering in ferromagnetic $CdCr_2Se_4$," *Solid State Commun.*, **6**, 747–70 (1968).

W. F. Edgell and N. Pauuwe, "Contact ion pairs in solution and infrared evidence for the structure of the decacarbonyldichromium dianion $(Cr_2—(CO)_{10})^{2-}$," *Chem. Commun.*, 284–5 (1969).

A. Mueller, E. J. Barna, and P. J. Hendra, "Position of the deformation frequencies (E) and (F_2) in the isoelectronic ions vandate (V), chromate (VI) and permanganate," *Spectrochim. Acta*, **25A**, 1654–6 (1969) (Ger.).

W. Krasser, "Vibrational spectrum and (valence) force fields of sodium-hexanitritocobaltate," *Z. Naturforsch, A*, **24**, 1667–8 (1969) (Ger.)

H. Seibert and H. H. Eysel "Vibrational spectroscopic study of hexaamminecobalt (III) and hexaamminechromium (III) salts," *J. Mol. Struct.*, **4**, 29–40 (1969) (Ger.).

COPPER AND NICKEL

M. Balkanski, M. Nusimovici, and J. Reydellet "First order Raman spectrum of Cu_2O," *Solid State Commun.*, **7**, 815–18 (1969).

J. R. Beattie, T. R. Gilson, and G. A. Ozin, "Single-crystal Raman spectroscopy of square-planar and tetrahedral $CuCl_4^{2-}$ ions, of the $ZnCl_4^{2-}$ ion, and of cupric chloride dihydrate," *J. Chem. Soc. A*, 534–41 (1969).

L. L. Garber, L. B. Sims, and C. H. Brubaker, Jr.," A vibrational analysis of the tetrazolate ion and the preparation of bis (tetrazolato)copper (II)," *J. Amer. Chem. Soc.*, **90**, 2518–23 (1968).

D. Jones, I. J. Hyams, and E. R. Lippincott, "Laser Raman spectrum and vibrational assignment of the tetracyanonickelate ion," *Spectrochim. Acta*, **24A**, 973–80 (1968).

L. H. Jones, R. S. McDowell, and M. Goldblatt, "Force constants of nickel carbonyl from vibrational spectra of isotopic species," *J. Chem. Phys.*, **48**, 2663–70 (1968).

M. J. Resifeld, "Absorption spectrum of potassium hexafluoronickelate. IV. Infrared and Raman spectra," *J. Mol. Spectrosc.*, **29**, 120–7 (1969).

R. T. Bailey, "Infrared and laser Raman spectra of methylcyclopentadienyl nickel nitrosyl," *Spectrochim. Acta*, **25A**, 1129–33 (1969).

GERMANIUM

K. M. Mackay and K. J. Sutton, "Isomers of pentagermane and tetragemane," *J. Chem. Soc. A*, 2312–16 (1968).

P. A. W. Dean, D. F. Evans, and R. F. Phillips, "Tri-oxalato complexes of silicon, germanium, tin, and titanium," *J. Chem. Soc. A*, 363–6 (1969).

R. J. Bell, N. F. Bird, and P. Dean, "The vibrational spectra of vitreous silica, germania, and beryllium fluoride," *Proc. Phys. Soc. London, Solid State Phys.*, **1**, 299–303 (1968).

S. Cradock and E. A. V. Ebsworth, "Germyl chemistry. IV. Germyl azide," *J. Chem. Soc. A*, 1420–3 (1968).

S. Cradock and E. A. V. Ebsworth, "Germyl Chemistry. VI. Digermylcarbodiimide," *J. Chem. Soc. A*, 1423–6 (1968).

J. E. Drake and C. Riddle, "Vibrational spectra of digermylarsine and digermylphosphine," *Inorg. Chem. Acta*, **3**, 136–8 (1969).

J. E. Drake, C. Riddle, and D. E. Rogers, "Vibrational spectra of germane and silane derivatives. I. Fundamental vibration frequencies of dichlorogermane," *J. Chem. Soc. A*, 910–13 (1969).

K. M. Mackay, K. J. Sutton, S. R. Stobart, J. E. Drake, and C. Riddle, "Vibrational spectra of monogermylphosphine, monogermylphosphine-d_2, monogermyl-d_3-phosphine, and monogermyl-d_3-phosphine-d_2," *Spectrochim. Acta*, **25A**, 925–40 (1969).

J. E. Drake, C. Riddle, K. M. Mackay, S. R. Stobart, and K. J. Sutton, "Vibrational spectra of monogermylarsine, monogermylarsine-d_2, monogermyl-d_3-arsine and monogermyl-d_3-arsine-d_2," *Spectrochim. Acta* **25A**, 941–51 (1969).

J. E. Drake and C. Riddle, "Vibrational spectra of germane and silane derivatives. II. Difluro-, dibromo-, and diiodo-germane," *J. Chem. Soc. A*, 14, 2114–17 (1969).

J. R. Durig, C. W. Sink, and J. B. Turner, "Vibrational spectra and structure of organogermanes. Low wavenumber vibrations of some triphenylgermanes," *Spectrochim. Acta*, **25A**, 629–45 (1969).

HALOGENS

A. Bree and R. Zwarich, "Vibrational assignment of fluorene from the infrared and Raman spectra," *J. Chem. Phys.*, **51**, 912–20 (1969).

M. Suzuki, T. Yokoyama, and M. Ito, "Raman spectrum and intermolecular forces of the chlorine crystal," *J. Chem. Phys.*, **50**, 3392–8 (1969).

J. E. Cahill and G. E. Leroi, "Raman spectra of solid chlorine and bromine," *U. S. Govt. Res. Develop. Rep.*, **69**, 65 (1969).

M. Suzuki, T. Yokoyama, and M. Ito, "Raman spectrum of the bromine crystal," *J. Chem. Phys.*, **51**, 1929–31 (1969).

P. M. Richardson and E. R. Nixon, "Lattice vibration of solid cyanogen," *J. Chem. Phys.*, **49**, 4276–84 (1968).

M. Pezolet and R. Savoie, "Raman spectra of liquid and crystalline hydrogen cyanide and deuterium cyanide," *Can. J. Chem.*, **47**, 3041–8 (1969).

K. Kawai, Y. Kodama, and F. Mizrikanni, "Infrared and Raman spectra of 2,4-dichloro-6-isocyanodichloro-s-triazine (tetrameric cyanogen chloride)," *Spectrochim. Acta*, **24A**, 1013–16 (1968).

T. Barrowcliffe, I. R. Beattie, P. Day, and K. Livingston, "The vibrational spectra of some chloro anions," *J. Chem. Soc. A*, 1810–12 (1967).

K. O. Christe, J. P. Guestin, and W. Sawodny, "Infrared and Raman spectra of the hexafluoroiodine (V) anion, IF_6^-," *Inorg. Chem.*, **7**, 626–8 (1968).

G. A. Olah and M. B. Comisarow, "Radical cations. II. Chlorine monofluoride molecule cation, ClF^+," *J. Amer. Chem. Soc.*, **91**, 2172–3 (1969).

K. O. Christe and W. Sawodny, "Vibrational spectra, force constants, and bonding of the tetrafluorochlorine (III) anion, ClF_4^-, and of chlorine pentafluoride," *Z. Anorg. Allg. Chem.*, **357**, 125–33.

J. Shamir and I. Yaroslavsky, "Laser Raman spectra of some halogen fluoride ions $Br F_4^-$, BrF_6^-, IF_4^+, and IF_4^-," *Isr. J. Chem.*, **7**, 495–7 (1969).

H. Selig and H. Holzman, "Raman spectrum of iodine pentafluoride: evidence for polymerization in the liquid state," *Isr. J. Chem.*, **7**, 417–20 (1969).

H. Stammreich and Y. Kawano, "Raman spectrum of solid iodine trichloride," *Spectrochim. Acta*, **24A**, 899–904 (1968).

H. H. Claassen, E. L. Gasner, and H. Selig, "Vibrational spectra of iodine heptafluoride and rhenium heptafluoride," *J. Chem. Phys.*, **49**, 1803–7 (1968).

H. Selig and E. L. Gasner, "The rhenium hepafluoride-hydrogen fluoride system," *J. Inorg. Nucl. Chem.*, **30**, 658–9 (1968).

P. J. Hendra and P. J. D. Park, "Vibrational spectra of sulfur and selenium nonohalides," *J. Chem. Soc. A*, 908–11 (1968).

D. C. Moule and C. R. Subramanian, "Vibrational spectra of thiocarbonyl chlorofluoride and a normal coordinate analysis of thiocarbonyl chlorofluoride and thiocarbonyl difluoride," *Can. J. Chem.*, **47**, 1011–17 (1969).

P. R. Reed and R. W. Lovejoy, "Vibrational spectrum of sulfuryl bromide fluoride," *Spectrochim. Acta*, **24A**, 1795–1801 (1968).

S. G. Frankiss, "Vibrational spectra and structures of S_2Cl_2, S_2Br_2, Se_2Cl_2 and Se_2Br_2," *J. Mol. Struct.*, **2**, 271–9 (1968).

E. B. Bradley, C. A. Frenzel, and M. J. Mathur, "Measurements of the Raman and the far-infrared spectrum of sulfur bromide," *J. Chem., Phys.*, **49**, 2344–6 (1968).

E. B. Bradley, M. S. Mathur, and C. A. Frenzel, "New measurements of the infrared and the Raman spectrum of S_2Cl_2," *J. Chem. Phys.*, **47**, 4325–9 (1967).

H. F. Shurvell and H. J. Bernstein, "Raman spectrum of solid sulfur hexafluoride," *J. Mol. Spectrosc.*, **30**, 153–7 (1969).

J. C. Evans and G. Y. S. Lo, "Vibrational spectra of the hydrogen dihalide ions. IV. $BrHBr^-$ and $BrDBr^-$," *J. Phys. Chem.*, **71**, 3942–7 (1967).

J. Shamir and J. Binenboym, "Laser Raman spectrum of the N_2F^+ cation," *J. Mol. Struct.*, **4**, 100–3 (1969).

Y. R. Shen, H. Rosen, and F. Stenman, "Frequency shift of the stretching vibration of iodine in liquid mixtures," *Chem. Phys. Lett.*, **1**, 671–4 (1968).

S. P. Beaton, D. W. A. Sharp, A. J. Perkins, I. Shift, H. H. Hyman, and K. Christe, "Vibrational spectra of hexafluorooiodate salts," *Inorg. Chem.*, **7**, 2174–6 (1968).

I. W. Levin, "Low-temperature Raman cell: solid spectra of the ν_1 vibrations of the Group IV tetrachlorides," *Spectrochim. Acta*, **25A**, 1157–60 (1969).

J. S. Avery, C. D. Burbridge, and D. M. L. Goodgame," "Raman spectra of tetrahalo anions of iron (III), manganese (II), iron (II), copper (II) and zinc (II)," *Spectrochim. Acta*, **24A**, 1721–6 (1968).

S. J. Shearer, G. C. Turrell, J. I. Bryant, and R. L. Brooks III, "Vibrational spectra of cyanuric triazide," *J. Chem. Phys.*, **48**, 1138–44 (1968).

S. Abramowitz and I. W. Levin, "Raman spectrum of ONF_3," *J. Chem. Phys.*, **51**, 463–4 (1969).

J. C. Carter, R. J. Bratton, and J. F. Jackovitz, "Structure and vibrational spectrum of nitrogen trichloride," *J. Chem. Phys.*, **49**, 3751–4 (1968).

P. J. Hendra and J. K. Mackenzie, "The laser Raman Spectrum of nitrogen trichloride," *Chem. Commun.* 760–2 (1968).

A. Mueller, A. Ruoff, B. Krebs, O. Glemser, and W. Koch, "Infrared, Raman, and electron absorption spectra, coriolis coupling constants, thermodynamic functions, and bonding ratios of thiazyl trifluoride, NSF_3," *Spectrochim. Acta*, **25A**, 199–205 (1969). (Ger.).

P. J. Hendra and F. Jovic, "The Raman spectra of some complex anions of the formula $(MX_6)^{2-}$ in the solid phase and in solution, where M = Se and Te; X = Cl, Br and I," *J. Chem., Soc. A*, 600–2 (1968).

W. J. Balfour and G. W. King, "Infrared and Raman spectra of oxalyl bromide," *J. Mol. Spectros.*, **28**, 411–14 (1968).

J. E. Cahill and G. E. Leroi, "Raman spectra of solid chlorine and bromine," *J. Chem. Phys.* **51**, 4514–19 (1969).

MERCURY COMPOUNDS

J. E. D. Davies and D. A. Long, "Vibrational spectra of the halides and complex halides of the Group IIB elements. II. Raman spectroscopic study of the systems HgX_2-LiX (X = Cl, Br) in aqueous and tributyl phosphate solutions," *J. Chem. Soc. A*, 2564–8 (1968).

C. W. Bradford, W. Van Bronswyk, R. J. H. Clark, and R. S. Nyholm, "Preparation and vibrational spectra of compounds of the type $M(CO)_4(HgX)_2$," *J. Chem. Soc. A*, 2456–63 (1968).

C. C. Addison, D. W. Amos, and D. Sutton, "Infrared and Raman spectra of solutions of zinc, cadium and mercury (II) nitrates in acetonitrile," *J. Chem. Soc. A.*, 2285–90 (1968).

R. P. J. Cooney, J. R. Hall, and M. A. Hooper "Raman spectra of mercury (I) and mercury (II) iodides in the solid state," *Aust. J. Chem.*, **21**, 2145–52 (1968).

A. J. Melveger, R. K. Khanna, B. R. Guscott, and E. R. Lippincott, "Low-frequency laser-excited Raman spectral study of the red to yellow phase transition in mercuric iodide," *Inorg. Chem.*, **7**, 1630–4 (1968).

J. S. Coleman, R. A. Penneman, L. H. Jones, and I. K. Kressin, "Mixed ligand complexes in mercury (II)-cyanide-iodide solutions: A Raman and ultraviolet study," *Inorg. Chem.*, **7**, 1174–6 (1968).

J. R. Saraf, R. C. Aggarwal, and J. Prasad, "Raman spectra of mixed trihalomercurate (II) complex anions," *J. Inorg. Nucl. Chem.*, **31**, 2123–6 (1969).

J. C. Evans " Infrared and Raman spectra of solutions of zinc, cadmium and mercury (II) nitrates in acetonitrile," *J. Chem. Soc. A*, 1849 (1969).

R. P. J. Cooney and J. R. Hall, "Analysis of the vibrational spectra of orthorhombic mercury (II) oxide and hexogonal mercury (II) sulfide," *Aust. J. Chem.*, **22**, 331–6 (1969).

R. P. J. Cooney and J. R. Hall, " Vibrational spectra of mercury (II) and methyl-mercury II thiocyanates," *Aust. J. Chem.*, **22**, 2117–23 (1969).

K. Krishman and R. A. Plane, " Raman and infrared spectra of *o*-phenanthroline and its complexes with zinc(II) and mercury(II)," *Spectrochim. Acta*, **25A**, 831–7 (1969).

A. Lowenschuss, A. Ron, and O. Schnepp, "Vibrational spectra of Group IIB halides of cadmium and mercury," *J. Chem. Phys.*, **50**, 2502–12 (1969).

J. H. R. Clarke and C. Solomons, "Raman spectra of mercuric iodide, iodochloride and iodobromide in the molten state," *J. Chem. Phys.*, **48**, 528–9 (1968).

J. R. Durig, K. K. Lau, G. Nagarajan, M. Walker, and J. Bragin, "Vibrational spectra and molecular potential fields of mercurous chloride, bromide, and iodide," *J. Chem. Phys.*, **50**, 2130 (1969).

MOLYBDENUM AND TUNGSTEN

C. G. Barlow, J. F. Nixon, and M. Webster, "Chemistry of phosphorus-fluorine compounds. IX. Preparation and spectroscopic studies of fluorophosphine-molybdenum carbonyl complexes," *J. Chem. Soc. A*, 2216–23 (1968).

D. Hartley and M. J. Ware, " Raman spectra and vibrational analysis of molybdenum cluster compounds," *Chem. Commun.*, 912–13 (1967).

A. Mueller, B. Krebs, R. Kababcioghi, M. Stockburger, and O. Glemser, " Vibrational spectra and force constants of tetraselenomolybdate (VI) and tetraseleno-tungstate (VI). Raman spectra of ammonium dithiomolybdate (VI) and ammonium dithiotungstate (VI)," *Spectrochim. Acta*, **24A**, 1831–7 (1968) (Ger.).

J. E. Fergusson, "Halide chemistry of chromium, molybdenum and tungsten," *Halogen Chem.*, **3**, 227–302 (1967).

R. K. Khanna, W. S. Brower, B. R. Guscott, and E. R. Lippincott, "Laser induced Raman spectra of some tungstates and molybdates," *J. Res. Nat. Bur. Stand.*, **A72**, 81–4 (1968).

R. V. Parish, P. G. Simms, M. A. Wells, and L. A. Woodward, "Eight-coordination. IV. Structures of the octacyanide complexes of quadri- and quinque-valent molybdenum and tungsten," *J. Chem. Soc. A*, 2882–6 (1968).

K. O. Hartman and F. A. Miller, "Raman and infrared spectra and structure of several eight-coordinated ions: octacyanomolybdate (IV), octacyanotungstate (IV), and octafluorotantalate (V)," *Spectrochim. Acta*, **24A**, 669–84 (1968).

H. J. Clase, A. M. Noble, and J. M. Winfield, "Nature of tungsten hexafluoride in solution," *Spectrochim. Acta*, **25A**, 293–5 (1969).

J. F. Scott, "Dipole-dipole interactions in tungstates," *J. Chem. Phys.*, **49**, 98–100 (1968).

R. A. Walton, "Laser Raman spectrum of crystalline tungsten (VI) chloride," *Chem. Commun.*, 1385 (1968).

D. M. Adams, G. W. Fraser, D. M. Morris, and R. O. Peacock, "Vibrational spectra of halides and complex halides. V. Tungsten chloride pentafluoride," *J. Chem. Soc. A*, 113–3 (1968).

NIOBIUM AND TANTALUM

G. A. Ozin, G. W. A. Fowles, D. J. Tidmarsh, and R. A. Walton, "Complex halides of transition metals. IX. Preparation and electronic vibrational spectra of the hexahaloniobates (V) and hexahalotantalates (V) $(Et_4N)MX_5Y$ ($X = Cl$ or Br, $Y = Cl$, Br, or I) including a vibrational analysis of the MX_5Y^- anions," *J. Chem. Soc. A*, 642–6 (1969).

R. D. Werder, R. A. Frey, and Hs. H. Guenthard, "Far infrared matrix and solution spectra and solid state vibrational spectra of niobium pentachloride," *J. Chem. Phys.*, **47**, 4159–65 (1967).

H. Selig, A. Reis, and E. L. Gasner, "Raman spectra of liquid niobium pentafluoride and tantalum pentafluoride," *J. Inorg. Nucl. Chem.*, **30**, 2087–90 (1968).

G. Burns, J. D. Axe, and D. F. O'Kane, "Raman measurements of $NaBa_2Nb_5O_{15}$ and related ferroelectrics," *Solid State Commun.*, **7**, 933–6 (1969).

C. H. Perry and N. E. Tornberg, "Optical phonons in mixed potassium tantalates," *Phys. Rev.*, **183**, 595–603 (1969).

C. H. Perry and N. E. Tornberg, "The Raman spectrum of lead titanate and solid solutions of sodium and potassium tantalates and potassium tantalates and niobates," *NASA-CR-96362*. Avail. CFSTI.

I. R. Beattie, T. E. Gilson, and G. A. Ozin, "Vibrational spectra of $Ca_3Tl_2Cl_2$, Au_2Cl_6, Nb_2Cl_{10}, Nb_2Br_{10}, Ta_2Cl_{10}, Ta_2Br_{10}," *J. Chem. Soc. A*, 2765–71 (1968).

F. J. Farrell, V. A. Maroni, and T. G. Spiro, "Vibrational analysis for octahedral oxyanions hexaniobate (V) and hexatantalate (V) and the Raman intensity criterion for metal-metal interaction," *Inorg. Chem.*, **8**, 2638–42 (1969).

OSMIUM AND RUTHENIUM

G. Davidson, N. Logan, and A. Morris, "Raman spectrum of ruthenium and osmium tetroxides," *Chem. Commun.*, **17**, 1044–6 (1968).

W. P. Griffith, "Raman spectra of ruthenium tetroxide and related species," *J. Chem. Soc. A*, 1663–4 (1968).

C. O. Quicksall and T. G. Spiro, "Raman frequencies of metal cluster compounds: triosmium dodecacarbonyl and triruthenium dodecacarbonyl," *Inorg. Chem.*, **7**, 2365–9 (1968).

I. W. Levin, "Low-temperature Raman spectra of solid osmium tetroxide and ruthenium tetroxide," *Inorg. Chem.*, **8**, 1018–21 (1969).

J. Chatt, A. B. Nikolsky, R. L. Richards, and J. R. Sanders, "Raman spectrum and structure of the ion $(Ru-(NH_3)_5)_2N_2)^{4+}$," *Chem. Commun.*, 154–5 (1969).

C. O. Quicksall and T. G. Spiro, "Raman frequencies of metal cluster compounds: $Os_3(CO)_{12}$ and $Ru_3(Co)_{12}$," *New Aspects Chem. Metal Carbonyls Deriv., Int. Symp. Proc.*, 1st F6/20 pp. (1968).

OXYGEN SPECIES

J. Selig and H. H. Classen, "Raman spectrum of ozone," *Israel J. Chem.*, **6**, 499–500 (1968).

J. Rolfe, W. Holzer, W. F. Murphy, and H. J. Bernstein, "Spectroscopic constants for O^{2-} ions in alkali halide crystals," *J. Chem. Phys.*, **49**, 963 (1968).

W. Holzer, W. F. Murphy, H. J. Bernstein, and J. Rolfe, "Raman spectrum of O^{2-} ion in alkali halide crystal," *J. Mol. Spectrosc.*, **26**, 543–5 (1968).

J. Shamir, J. Binenboym, and H. H. Claassen, "The vibrational frequency of the oxygen molecule $(O_2{}^+)$ cation," *J. Amer. Chem. Soc.*, **90**, 6223–4 (1968).

J. L. Arnau and P. A. Giguere, "Spectroscopic study of hydrogen peroxide derivatives. I. Hyperol, $Co(NH_2)_2H_2O_2$," *J. Mol. Struct.*, **3**, 483–9 (1969) (Fr.).

INORGANIC COMPOUNDS AND PHOSPHORUS COMPOUNDS

A. C. Chapman, "Spectra of phosphorus compounds. III. Vibrational assignments and force constants of P_4O_6 and P_4O_{10}," *Spectrochim. Acta*, **24A**, 1687–96 (1968).

I. W. Levin, "Raman spectrum of solid phosphorus pentafluoride," *J. Chem. Phys.*, **50**, 1031 (1969).

E. Mayer and R. E. Hester, "Infrared, Raman, and nuclear magnetic resonance spectra of isotopically substituted bisboranohypophosphite anions," *Spectrochim. Acta*, **25A**, 237–43 (1969).

J. J. Rush, A. J. Melveger, T. C. Farrar, and T. Tsang, "Laser-Raman spectra and hindered rotation in the phosphonium halides," *Chem. Phys. Lett.*, **2**, 621–4 (1968).

J. R. Durig, D. J. Antion, and F. G. Baglin, "Far-infrared and Raman spectra of phosphonium iodide and phosphomium iodide-d_4," *J. Chem. Phys.*, **49**, 666–74 (1968).

S. G. Frankiss, "Vibrational spectrum and structure of solid diphosphine," *Inorg. Chem.*, **7**, 1931–3 (1968).

L. C. Kravitz, J. D. Kingsley, and E. K. Elkin, "Raman and infrared studies of coupled phosphate vibrations," *J. Chem. Phys.*, **49**, 4600–10 (1968).

W. Yellin and W. A. Gilley, "Vibrational spectrum of tin (II) orthophosphate $SnHPO_4\cdot\frac{1}{2}H_2O$," *Spectrochim. Acta*, **25A**, 879–87 (1969).

R. W. Mitchell, L. J. Kuzma, R. J. Pirkle, and J. A. Merritt, "Vibrational spectra of phosphirane, phosphirane-1-d_1, and phosphirane-2,2,3,3,-d_4," *Spectrochim. Acta*, **25A**, 819–29 (1969).

G. E. Coxon and D. B. Sowerby, "Cyclic inorganic compounds. VI. Vibrational spectra of pentameric phosphonitrile bromide," *Spectrochim. Acta*, **24A**, 2145–9 (1968).

I. R. Beattie, K. M. S. Livingston, and D. J. Reynolds, "Vibrational spectra of phosphorus chloride$_n$-fluoride$_{5-n}$," *J. Chem. Phys.*, **51**, 4269–71 (1969).

J. R. Durig, D. J. Antion, and C. B. Pate, "Far infrared and Raman spectra of phosphonium bromide and phosphonium bromide—d_4," *J. Chem. Phys.*, **51**, 4449–56 (1969).

J. J. Rush, A. J. Melvager, and E. R. Lippincott "Laser Raman spectra of phosphonium iodide, phosphonium bromide and phosphonium chloride, *J. Chem. Phys.*, **51**, 2947–55 (1969).

H. F. Shurwell, "Vibrational assignment of thiophosphorylfluoride," *Spectrochim. Acta*, **25A**, 973–6 (1969).

1. C. Hisatsune, "Assignment of the 2,2,4,4,6,6,8,8-octachloro-1,3,5,7-tetraazo-2,4,6,8-tetraphosphocine vibrational spectrum," *Spectrochim. Acta*, **25A**, 301–12 (1969).

T. R. Manly and D. A. Williams, "Vibrational assignments of some phosphonitrilic halides," *Spectrochim. Acta*, **24A**, 1661–2 (1968).

T. J. Kistenmacher and G. D. Stucky, "Structural and spectroscopic studies of tetrachlorophosphonium-tetrachloroferrate (III)," *Inorg. Chem.*, **7**, 2150–5 (1968).

J. H. S. Green, "Vibrational spectra of ligands and complexes. Triethylphosphine and some related compounds," *Spectrochim. Acta*, **24A**, 137–43 (1968).

R. A. Nyguist and W. W. Muelder," Infrared and Raman study of O,O-dimethyl and O,O-dimethyl-d_6-phosphorochloridothioate," *Spectrochim. Acta*, **24A**, 187–201 (1968).

J. R. Durig, M. Walker, and F. G. Baglin, "Lattice vibrations of molecular crystals. Cyanamide and cyanamide-d_2," *J. Chem. Phys.*, **48**, 4675–82 (1968).

PLATINUM AND PALLADIUM

R. G. Denning and M. J. Ware, "Vibrational spectra of platinum (II) complexes. I. Raman and infrared spectra of the $(Pt(CO)Cl_3)$ and $(Pt(NH_3)Cl_3)$ ions," *Spectrochim. Acta*, **24A**, 1785–93 (1968).

M. J. Taylor, A. L. Odell, and H. A. Raethel, "Assignment of the stretching frequencies in some chlorine-bridged platinum (II) complexes $Pt_2Cl_4L_2$, with phosphorus and arsenic ligands," *Spectrochim. Acta*, **24A**, 1855–61 (1968).

J. R. Allkins and P. J. Hendra, "Vibrational spectra of coordination compounds of formula trans MX_2Y. IV. Effect of coordination on the internal modes when X = chloro, bromo or iodo, M = palladium or platinum and Y = dimethyl sulfide, selenide, or telluride," *Spectrochim. Acta*, **24A**, 1305–10 (1968).

D. W. James and M. J. Nolan, "Vibrational spectra of the hexachloroplatinate (IV) anion," *Inorg. Nucl. Chem. Lett.*, **4**, 97–9 (1968).

P. A. Bulliner and T. G. Spiro, "Platinum-oxygen stretching and hydroxyl wagging frequencies in trimethylplatinum hydroxide," *Inorg. Chem.*, **8**, 1023–5 (1969).

INORGANIC COMPOUNDS—SULFUR

A. Anderson and Y. T. Loh, "Low-temperature Raman structure of rhombic sulfur," *Can. J. Chem.*, **47**, 879–84 (1969).

I. R. Beattie, M. J. Gall, and G. A. Ozin, "Single-crystal Raman studies and the vibrational spectrum of the dithionate ion," *J. Chem. Soc. A*, 1001–8 (1969).

W. C. Holton and M. DeWit, "Raman spectrum of (sulfur and selenium molecular ions) S_2^- and Se_2^- in potassium iodide," *Solid State Commun.*, **7**, 1099–101 (1969).

A. T. Ward, "Raman spectroscopy of sulfur, sulfur-selenium, and sulfur-arsenic mixtures," *J. Phys. Chem.*, **72**, 4133–9 (1968).

A. T. Ward, "Raman spectrum and force constants of S_4^{2-}," *Mater. Res. Bull.*, **4**, 581–90 (1969).

R. E. Miller and G. E. Leroi, "Raman spectra of polycrystalline hydrogen sulfide and deuterium sulfide," *J. Chem. Phys.*, **49**, 2789–97 (1968).

H. Wieser, P. J. Krueger, E. Muller, and J. B. Hyne, "Vibrational spectra and a force field for H_2S_3 and H_2S_4," *Can. J. Chem.*, **47**, 1633–7 (1969).

V. Ananthanarayanan, "The vibrational spectrum of the sulfate ion in crystalline $M'M''(SO_4)_2 6H_2O$ (M' = potassium or ammonium and M'' = magnesium, zinc, nickel or cobalt. Observations on the symmetry of the sulfate ion in crystals," *J. Chem. Phys.*, **48**, 573–81 (1968).

N. Krishnamurthy and R. S. Katiyar, "Raman and infrared spectra of sodium ethyl sulfate monohydrate," *Indian J. Pure Appl. Phys.*, **7**, 97–9 (1969).

E. R. Clark and A. J. Collett, "Infrared and Raman spectra of seleno-and telluro-pentathionates," *J. Chem. Soc. A*, 1594–6 (1969).

N. Kranzman and M. Kranzman, "Vibrational spectra of potassium dithionate," *C. R. Acad. Sci., Ser. B.*, **269**, 641–3 (1969) (Fr.).

THALLIUM

A. J. Carty, "Metal-halogen stretching vibrations in coordination complexes of gallium, indium, and thallium," *Coord. Chem. Rev.*, **4**, 29–39 (1969).

H. Stammreich, B. M. Chadwick, and S. G. Frankiss, "Vibrational spectrum and structure of solid thallium(I)dicyanoaurate (I)," *J. Mol. Struct.*, **1**, 191–6 (1968).

J. E. D. Davies and D. A. Long, "Vibrational spectra of the halides, complex halides, and mixed halides of thallium (III). I. The vibrational spectra of $TlBr_3$, $TlBr_4$, $TlCl_6^{-3}$ and $TlCl_4^{-}$," *J. Chem. Soc. A.*, 2050–4 (1968).

D. M. Adams and D. M. Morris, "Vibrational spectra of halides and complex halides. IV. Some tetrahalothallates and the effects of *d*-electronic structure on the frequencies of hexachlorometallates," *J. Chem. Soc., A*, 6945 (1968).

TIN

I. Wharf and D. F. Shriver, "Vibrational frequencies and intramolecular forces in anionic tin-halogen complexes and related species," *Inorg. Chem.*, **8**, 914–25 (1969).

K. G. Huggins, F. W. Parrett, and H. A. Patel, "Vibrational spectra of some adducts of tin tetraiodide," *J. Inorg. Nucl. Chem.*, **31**, 1209–12 (1969).

TITANIUM

D. M. Adams and D. C. Newton, "Vibrational spectra of halides and complex halides. The ions $(MCl_6)^{2-}$, M = titanium, zirconium, hafnium," *J. Chem. Soc. A*, 2262–3 (1968).

J. E. D. Davies and D. A. Long, "Vibrational spectra of titanium tetrachloride-hydrochloric acid and titanium tetrachloride-tributyl phosphate systems and the hexachloro anions of zirconium (IV), hafnium (IV) niobium (V) and tantalum (V)," *J. Chem. Soc. A*, 2560–4 (1968).

R. E. Collins "Acetylacetonatotitanium (IV) complex of composition TiOCl·$(C_5H_7O_2)$," *J. Chem. Soc. A*, 1895–7 (1969).

J. E. Griffiths, "Molecular association in titanium tetrachloride: laser Raman spectroscopy and chlorine isotope effects," *J. Chem. Phys.*, **49**, 642–7 (1968).

OTHERS

T. C. Damen, A. Kielan, S. P. S. Porto, and S. Singh, "Raman effect of cerous chloride and praesodymium chloride," *Solid State Commun.*, **6**, 671–3 (1968).

D. G. Karraker, "Raman and infrared spectra of ammonium hexanitrocerate (IV)," *Inorg. Nucl. Chem. Lett.*, **4**, 309–13 (1968).

D. F. Koster, "The vibrational spectra and structure of trichloroisocyanatosilane, trichlorothioisocyanatosilane, chlorotriisocyanatosilane, and *tert*-butylisocynate," *Spectrochim. Acta*, **24A**, 395–405 (1968).

D. M. Adams and D. M. Morris, "Vibrational spectra of halides and complex halides. III. Hexahalotellurates," *J. Chem. Soc. A*, 2067–9. (1967)

R. J. Gillespit and G. P. Pez, "Infrared and Raman spectra of the Se_4^{2+} ion," *Inorg. Chem.*, **8**, 1229–33 (1969).

G. W. King, K. Srikameswaran, "Vibrational spectrum of carbon diselenide," *J. Mol. Spectrosc.*, **29**, 491–4 (1969).

J. R. Durig, A. L. Marston, R. B. King, and L. W. Houk, "Infrared and Raman spectra of cyclopentadienylvanadiumtetracarbonyl derivatives: evaluation of the carbon-oxygen force constants," *J. Organometal. Chem.*, **16**, 425–37 (1969).

J. T. Kenney and F. X. Powell, "Raman spectra of molten manganese dichloride thallium trichloride, and tin dibromide," *U. S. Govt. Res. Develop. Rep.*, **69**, 56 (1969).

H. G. M. Edwards, M. J. Ware, and L. A. Woodward, "Vibrational spectra and stretching force-constants of tetrahalo complexes of manganese (II)," *Chem. Commun.*, 540–1 (1968).

L. A. Woodward and M. J. Ware, "Vibrational spectra of hexachlorouranate and herachlorothorate ions," *Spectrochim. Acta*, **24A**, 921–5 (1968).

E. L. Gasner and B. Frlec, "Raman spectrum of neptunium hexafluoride," *J. Chem. Phys.*, **49**, 5135–7 (1968).

R. E. Miller, K. L. Treuil, and G. E. Leroi, "Raman spectra of xenon difluoride-iridium pentafluoride and xenon difluoride-iodine pentafluoride adducts," *U. S. Govt. Res. Develop. Rep.*, **68**, 60–1 (1968).

H. H. Claassen, E. L. Gasner, H. Kim, and J. L. Huston, "Vibrational spectra and structure of XeO_2F_2," *J. Chem. Phys.*, **49**, 253–7 (1968).

F. O. Sladky, P. A. Bulliner, and N. Bartlett, "Xenon difluoride as a fluoride ion donor. Evidence for the salts $(Xe_2F_3)^+(MF_6)^-$, $(XeF)^+(MF_6)^-$, and $(XeF)^+ (M_2F_{11})^-$," *J. Chem. Soc. A*, 2179–88 (1969).

R. E. Hester and C. W. J. Scaife, "Vibrational spectra of molten salts. II. Infrared and Raman spectra of variably hydrated zinc nitrate," *J. Chem. Phys.*, **47**, 5253–8 (1967).

R. E. Hester and K. Krishnan, "Vibrational spectra of molten salts. V. Infrared and Raman spectra of some molten sulfates," *J. Chem. Phys.*, **49**, 4356–60 (1968).

W. G. Nilsen, "Raman spectrum of cubic zinc sulfide," *Phys. Rev.*, **182**, 838–50 (1969).

J. E. Griffiths, "Hexachlorodisilane: vibrational spectrum, structure, and internal rotation," *Spectrochim. Acta*, **25A**, 965–72 (1969).

B. J. Aylett and J. M. Campbell "Silicon-transition metal compounds. II. Preparation and properties of silylpentacarbonylmanganese," *J. Chem. Soc. A*, 1916–20 (1969).

N. A. Matwiyoff and W. G. Morius, "Proton magnetic resonance and Raman spectral studies of the complexes tetrakis (dimethylformamide)beryllium (II) and acetylacetonatobis (dimethyformamide)beryllium (II) in the solvent N, N-dimethylformamide. Direct determination of solvation numbers and kinetics of solvent exchange," *J. Amer. Chem. Soc.*, **89**, 6077–81 (1967).

I. J. Hyams, D. Jones and E. R. Lippincott, "The vibrational spectra of rhenium carbonyl and rhenium carbonyl iodide," *J. Chem. Soc. A*, 1987–93 (1967).

R. S. Katiyar and N. Krishnamurthy, "Raman and infrared spectra of beryllium sulfate tetrahydrate," *Indian J. Pure Appl. Phys.*, **7**, 95–7 (1969).

J. F. Scott, "Raman spectra of barium chlorofluoride, barium bromofluoride, and strontium chlorofluoride," *J. Chem. Phys.*, **49**, 2766–9 (1968).

H. C. Clark, K. R. Dixon, and J. G. Nicolson, "Preparation and properties of penta-flurosilicates," *Inorg. Chem.*, **8**, 450–3 (1969).

J. E. Durig and K. L. Hellams, "Low frequency vibrations of some organotrichloro-silanes," *Appl. Spectrosc.*, **22**, 153–60 (1968).

J. I. Bullock, "Raman and infrared spectroscopic studies of the uranyl ion: the symmetric stretching frequency, force constants, and bond lengths," *J. Chem. Soc. A.*, 781–4 (1969).

R. R. Berrett, B. W. Fitzsimmons, P. Gans, H. M. N. H. Irving, and P. Stratton, "Vibrational spectra and force-constant computations of cis- and trans-dicyano-notetrakis(methylisocyanide)iron(II)," *J. Chem. Soc. A.*, 904–8 (1969).

J. R. Durig and D. J. Antion, "Low frequency vibrations in ammonium iodide and ammonium bromide," *J. Chem. Phys.*, **51**, 3639–47 (1969).

G. C. Kulasingam, W. R. McWhinnie, and J. D. Miller, "Mechanism of formation and the stereochemistry of mono- and biscomplexes of rhodium (III) with 1,10-phenanthroline," *J. Chem. Soc. A*, 521–4 (1969).

E. W. Abel, R. A. N. McLean, "Vibrational spectra of dodecacarbonyltetrarhodium," *Inorg. Nucl. Chem. Lett.*, **5**, 381–4 (1969).

C. K. Asawa, "Raman spectrum of lanthanum tribromide," *Phys. Rev.*, **173**, 869–72 (1968).

E. R. Clark and A. J. Collett, "Preparation, infrared and Raman spectra of some compounds of tellurium (II)," *J. Chem. Soc. A*, 2129–30 (1969).

T. G. Spiro, V. A. Maroni, and C. O. Quicksall, "Revised 'Cluster' Raman frequencies for $Pb_6O-(OH)_6^{+4}$," *Inorg. Chem.*, **8**, 2524–6 (1969).

P. Cailet and P. Saumagne, "Infrared and Raman spectroscopic study of anhydrous alkali monomolybdates and monotungstates," *J. Mol. Struct.*, **4**, 191–201 (1969) (Fr.).

R. K. Khanna, C. W. Brown and L. H. Jones, "Laser Raman spectra of a single crystal of sodium nitroprusside and the vibrational frequencies of the $Fe(CN)_5 -NO^{2-}$ ion." *Inorg. Chem.*, **8**, 2195–200 (1969).

J. E. Griffiths and D. F. Sturman, "Hexachlorodisiloxane. Vibrational spectrum and structure," *Spectrochim. Acta*, **25A**, 1415–22 (1969).

J. R. Durig, W. H. Green, and A. L. Marston, "Low frequency vibrations of molecular crystals. IV. Boric acid," *J. Mol. Struct.*, **2**, 19–37 (1968).

S. A. Miller, H. H. Caspers, and H. E. Rast, "Lattice vibrations of yttrium vanadate," *Phys. Rev.*, **168**, 964–9 (1968).

P. J. Hendra, "The Vibrational spectrum of the permanganate ion," *Spectrochim. Acta*, **24A**, 125–9 (1968).

AUTHOR INDEX

167

SUBJECT INDEX

169